# SUPER VOLCANO

# SUPER VOLCANO

## The Ticking Time Bomb Beneath

## Yellowstone National Park

GREG BREINING

VOYAGEUR
PRESS

# DEDICATION

Books are born of inspiration—not necessarily one's own. I'd like to thank Michael Dregni for suggesting I write about the Yellowstone volcano, and Danielle J. Ibister for her meticulous attention to the manuscript. Thanks also to Paul Doss for his inspired, and inspirational, introduction to the park's geology, and to Mike Voorhies for his sense of wonderment at a living scene long passed from the face of the earth. Finally, I thank my wife, Susan, for supporting my endeavors.

The bright sun was extinguish'd, and the stars
Did wander darkling in the eternal space,
Rayless, and pathless, and the icy earth
Swung blind and blackening in the moonless air;
Morn came and went—and came, and brought no day.

—Lord Byron, *Darkness*

First published in 2007 by Voyageur Press, an imprint of The Quarto Group, 100 Cummings Center, Suite 265D, Beverly, MA 01915 USA. T (978) 282-9590 F (978) 283-2742 www.QuartoKnows.com

Voyageur Press titles are also available at discount for retail, wholesale, promotional, and bulk purchase. For details, contact the Special Sales Manager by email at specialsales@quarto.com or by mail at The Quarto Group, Attn: Special Sales Manager, 100 Cummings Center, Suite 265D, Beverly, MA 01915 USA.

10 9

Library of Congress Cataloging-in-Publication Data

Breining, Greg.
  Super volcano : the ticking time bomb beneath Yellowstone National Park / by Greg Breining.
      p. cm.
    ISBN: 978-0-7603-2925-2 (hardbound w/ jacket)

      1. Supervolcanoes—Yellowstone National Park. 2. Volcanoes—Yellowstone National Park. 3. Geology—Yellowstone National Park. I. Title.

  QE524.B74 2007
  551.2109787'5—dc22

2007014003

Edited by Danielle J. Ibister
Designed by Christopher Fayers

Front cover photograph by Frank Lukasseck, Corbis
Maps by Patti Isaacs, Parrot Graphics

Printed in the United States

# Contents

# Chapter 1

# The Big Blast

Yellowstone had been trembling and thundering for months, years, even centuries. The quakes came in swarms. There were hundreds each day, many too faint to feel. But often the earth shuddered. Faults broke through the surface, destroying trees, throwing clods of earth.

For years, the ground had been heaving upward. A region the size of Delaware, some forty miles by sixty miles centered on the southwestern corner of Yellowstone, had not only been shaking and splitting. It had also been rising and falling as though a giant deep below the ground was breathing, turning, heaving, sighing. And recently, all of this commotion had intensified. The ground was mostly rising now, as though something below, fierce and monstrous, was about to break loose.

As the ground shuddered, swelled, and stretched, fissures spread across the surface. Red-hot lava issued from deep within the earth. The heat and smoke were intense. At first the lava oozed and spread, setting the ground ablaze as it advanced. Then along the perimeter of this huge area, fissures spread and lava began to explode, heaving great slabs of molten rock.

Then something unimaginable happened, something on a scale that had rarely occurred before or since. As magma jetted to the surface, the release of pressure reached a tipping point. Gas contained in the pressurized magma boiled away violently. As magma escaped to the surface, the pressure dropped even more, and ever more vigorously, the gas escaped. It didn't fizz. Or roar. Or even merely blow up. It propagated exponentially with the speed and power of a nuclear explosion.

Earth-shaking eruptions occurred from three primary sites—what are now Island Park, Lewis Lake, and Old Faithful. In term of energy, imagine a machine gun burst issuing from Yellowstone, each blast the equivalent of a nuclear bomb. The magma chamber beneath the vast area of Yellowstone emptied with the force of hundreds of thousands of bombs the size of the explosion that leveled Hiroshima.

Until this moment, the Gallatin Range ran continuously into the Red Mountains, forming a spine through Yellowstone, clear south to the Tetons. In moments, or hours, or at the most, days—thirty miles of the Gallatins were obliterated. The continual explosions blew major portions of the mountains into the upper atmosphere and ejected six hundred cubic miles of magma. Imagine a block of rock more than eight miles by eight miles at the base, and more than eight miles high—a mountain far more massive than Everest—ejected from the earth, rocketed more than twenty miles into the stratosphere, blasted down the mountainsides, pulverized, melted, and reformed as rock. Water released from erupting magma billowed high into the atmosphere. Lightning storms thundered in the plumes of smoke, vapor, and ash.

---

*Imagine a block of rock more than eight miles by eight miles at the base, and more than eight miles high—a mountain far more massive than Everest—ejected from the earth.*

---

Glowing ash shot upward from each of the vents and blasted outward. These pyroclastic surges of hot gases and nearly vaporized particles of rock and volcanic glass raced down slopes and across the landscape as fast as several hundreds of miles per hour. Following close behind were pyroclastic flows—glowing, rushing rivers of heavier material, exceeding 1,000 degrees Fahrenheit. The hot flows uprooted, leveled, and burned whole forests and killed all animals from insects to bison.

Pyroclastic flows radiated outward from the volcano vents, running northward through the valleys of the Madison, Gallatin, and Yellowstone rivers, southward through the valley of the Snake, and broadly, as a sheet, southwest through the Snake River Plain. These cascades of glowing ash traveled as far as eighty miles before settling and welding to the ash beneath to form solid rock. This welded tuff ranged from 500 to 2,500 feet thick and covered 6,000 square miles, an area the size of Connecticut.

For hours, or days, explosions from the three vents continued to hurl gas, water vapor, and rock. Eventually, the magma chamber beneath Yellowstone emptied enough that the ground collapsed. The surface—what hadn't been vaporized and blown to bits—broke along the ring fissures at the edge of the immense chamber and dropped like a piston hundreds of yards into the vacated chamber below. Even massive Mount Washburn, a much older and now extinct volcano, sheared in half, the southern portion of the 10,000-foot mountain plunging into the evacuated magma chamber. This falling mass expelled more magma, which exploded from the fractures at the edges. The surrounding land, covered by ash and welded tuff, cascaded as avalanches into this newly formed caldera, more than fifty miles across. Much of the Gallatin Range fell into the fiery earth. For miles and miles all around, not a living thing remained. There was only a lunar landscape of hot rock, mud, and ash.

Dark clouds of ash and volcanic glass, blown throughout the stratosphere, drifted on the westerlies. The coarsest particles fell out first, as a light gray ash fall in perpetual darkness, burying areas near Yellowstone twenty feet deep. Finer bits of rock and glass remained aloft for hours or days, traveling hundreds of miles in the jet stream and lower-altitude winds. As the cloud blew eastward, the ash fell in decreasing depths and increasing fineness. Within days, snowlike ash fall blanketed the western United States from the Pacific to the present course of the Mississippi. Prairie winds blew the ash into drifts that filled swales to thirty feet deep. The finest particles of ash remained aloft for weeks.

Clouds of water vapor, carbon dioxide, hydrogen chloride, and hydrogen fluoride, ejected high into the atmosphere by the eruption, remained aloft for much longer. Sulfur dioxide converted to sulfuric acid, which

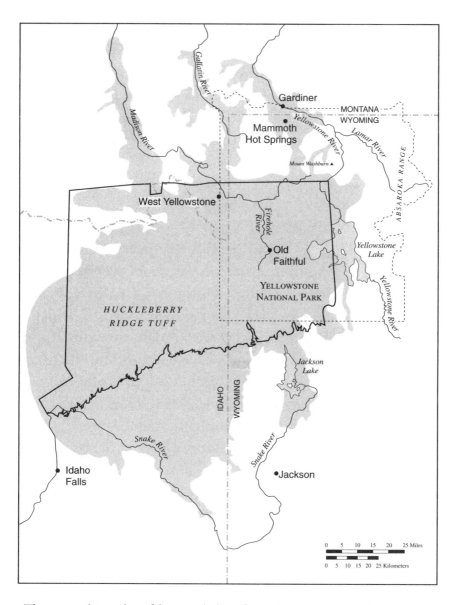

The 240 cubic miles of lava exploding from the magma chamber beneath Yellowstone 2.1 million years ago covered an area larger than the state of Connecticut (outline) with the so-called Huckleberry Ridge Tuff, resolidified volcanic rock. *Adapted from Robert L. Christiansen, "The Quaternary and Pliocene Yellowstone Plateau Volcanic Field of Wyoming, Idaho, and Montana"*

condensed to form sulfate aerosols. A vast cloud of these aerosols in the middle and lower stratosphere encircled the globe within weeks. Within a year, these aerosols completely covered the globe, reflecting the sun's radiation, causing the lower atmosphere to cool. For years the climate dropped several degrees, as the earth plunged into a long volcanic winter.

— . —

America's oldest and most famous national park sits atop one of the largest, most explosive volcanoes ever to exist.

We can only imagine the explosion at what we now call Yellowstone National Park. No one witnessed it. It happened 2.1 million years ago. We are left to estimate its power by mapping exposures of welded tuff and studying strata of ash beds in far-flung reaches of the West.

But scientists have concluded that the eruption was immense, the energy released unfathomable—2,500 times greater than the explosion that ripped Mount St. Helens in 1980, blowing off 1,300 feet of mountain, killing fifty-seven, and destroying forests for miles around eastern Washington. The Yellowstone eruption ejected more than a hundred times as much magma and ash as did Krakatau in Indonesia in 1883, when more than 36,000 died as a result of one of the most famous volcanic eruptions ever. Yellowstone—and a few other known volcanic eruptions—are termed "super volcanoes." That means they have expelled at least 1,000 cubic kilometers (about 240 cubic miles) of magma more or less all at once. No volcano in historic times, when literate humans were around to witness it, has even approached such magnitude.

But humans may get a chance to see Yellowstone blow firsthand. Because Yellowstone is not only the site of one of the largest volcanic eruptions ever known. It is also the largest, potentially most explosive, most violent, most deadly *active* volcano on the planet. Scientists say the chance of another catastrophic explosion someday is almost inevitable.

Since the cataclysmic explosion of 2.1 million years ago, two other large eruptions have rocked Yellowstone, buried it in volcanic rock, and left behind huge calderas. Though smaller than the first eruption, both subsequent explosions dwarfed better-known volcanoes, such as Mount St. Helens and Vesuvius. An explosion 1.3 million years ago blew 67 cubic miles of rock and ash across nearly 1,000 square miles, leaving behind a caldera 10 miles across at Island Park. An even bigger eruption 640,000 years ago spewed 240 cubic miles of lava and ash across 1,700 square miles, an area larger than Rhode Island. The caldera created by this most recent super explosion measures 50 miles by 30 miles and sits nearly in the center of Yellowstone National Park. Nearly a third of the park's 2.2 million acres either blew sky high or collapsed into the earth. The caldera is so filled with lava from subsequent smaller eruptions that most visitors never recognize they are standing on the remnants of an eruption so powerful it covered much of the continent with volcanic ash and transformed the earth's climate for years, or perhaps decades.

But the Yellowstone volcano didn't stop then. During the past 640,000 years, the caldera and surrounding areas have filled with lava from about eighty smaller eruptions. Some of the lava fields would have buried a thirty-story skyscraper. The most recent eruption occurred 70,000 years ago, but there's no reason to believe that the Yellowstone volcano is tapped out—and plenty of reason to believe it will reawaken.

Robert Christiansen, recently retired from the U.S. Geological Survey, is the geologist perhaps most responsible for piecing together the volcanic history of Yellowstone. He recently told the British Broadcasting Corporation: "Millions of people come to Yellowstone every year to see the marvelous scenery and the wildlife and all, and yet it's clear that very few of them really understand that they're here on a sleeping giant."

*Even today, the thin, heated crust of Yellowstone is heaving like the chest of a gasping man.*

And the giant sleeps fitfully, at that. What are the signs?

For one, the ground beneath Yellowstone is restless. About two thousand earthquakes shake the park each year—an average of more than five a day. Most are too small to feel, but some are significant. The Hebgen Lake quake of 1959, a 7.5 magnitude temblor centered just west of the park, opened a crack in the earth and triggered a landslide. Altogether, twenty-eight people died. The Borah Peak earthquake of 1983, centered fifteen miles west of Mackay, Idaho, registered nearly as strong—magnitude 7.3. A quake of 6.5 magnitude shook the Norris Geyser Basin in 1975. Ten years later, in October 1985, an earthquake swarm began to rock the park. For the next several months, more than three thousand quakes occurred; as many as two hundred shot through the ground each day. Ten years later, another swarm rattled the park like a nest of mad hornets.

For another, heat from magma near the surface drives the gushing geysers, blurping mud pots, and steaming fumaroles that distinguish the national park. The heat flow from the Firehole area, the location of Old Faithful, is seven hundred times the global average. Beneath the entire Yellowstone caldera (from the most recent super explosion), the average heat flow is thirty to forty times greater than the average worldwide, suggesting that new magma continues to heat the ground, perhaps maintaining another partially melted magma chamber. In fact, Christiansen believes the reservoir of magma is as much as ten times greater than what has already been expelled, as if the volcano is primed and ready. This reservoir of heat sustains the greatest concentration of geysers in the world. Steam and hot water have at times found violent and spectacular release, blowing out craters hundred of yards across during the last several thousand years.

Even today, the thin, heated crust of Yellowstone is heaving like the chest of a gasping man. The ground beneath Yellowstone is literally rising and falling. From 1923, when road surveyors determined the exact altitude of specific Yellowstone landmarks, until 1984, the ground beneath the central portion of the park bulged upward more than a yard. The breathing paused for a year or two. Then it began to drop. The land began rising again in 1996. As the ground north of Yellowstone Lake has bulged upward, the lake has inundated trees at the southern shore. In recent years, a global positioning system station at Steamboat, in the Norris Geyser Basin, has risen six inches a year. What causes the land to heave? Scientists say it may be superheated water and steam, or it may be magma. Either way, the source of heat is the same molten rock that could cause another eruption

So, hell yes, Yellowstone is still active. It's the lair of a fire-breathing dragon. Writes Robert Smith of the University of Utah, another giant of Yellowstone geologic research, "Yellowstone really is a living, breathing thing."

Says Christiansen in the pages of a comprehensive geologic report on the park—the kind of thing you'd think would be written cautiously: "Volcanism almost certainly will recur in the Yellowstone National Park region."

— · —

When such an explosion occurs, it will be a catastrophe. Christiansen calls it a potential "major human disaster"— not just for the park but for the entire region. The park itself will be destroyed, the forest leveled, and every living thing within vaporized, incinerated, blown to bits, buried in welded ash and mud, or asphyxiated by carbon dioxide. People in the surrounding towns of West Yellowstone, Gardiner, and Cooke City will face the same fate. Towns and cities throughout the West will have to dig out from deep volcanic ash. Cars, buses, planes will all be grounded till the ash clears. Respiratory illness will soon take a toll. Domestic livestock and wildlife west of the Mississippi will be stuck, and probably doomed. Great Plains wheat fields, the global breadbasket, will miss a season—or several—leading to widespread famine. The global volcanic winter will last years and influence the climate for decades or centuries. The worldwide death toll, says Michael Rampino, New York University associate professor of earth and environmental sciences, could reach a billion. "How do you get food, how do you get supplies, how do you get in and out, even after the eruption?" says Rampino, who has studied the climatic effects of huge volcanoes. "It depends on what you mean by killed—killed right away, starved to death?"

So we have something to look forward to—a probable disaster of global proportions. The question is when.

What is tantalizing—and a bit alarming—to consider is the timing of Yellowstone's super eruptions: 2.1 million years, 1.3 million years, 640,000 years ago. The intervals are 800,000 and 660,000 years. That suggests another explosion is due—and, in geologic time, soon!

Meanwhile, the geysers vent, the hot springs bubble, and the fumaroles hiss, powered by the magma beneath the ground, by the same magma that has charged the largest active volcano on earth and one of the largest volcanic eruptions ever to rock the planet. They remind us that the monster still stirs. How does it work? What are the chances it will reawaken? And what can we do when it does?

# CHAPTER 2

# YELLOWSTONE TODAY

A small group of us—adult students at a three-day class in Yellowstone National Park—gathered near the base of a 8,564-foot bump called Bunsen Peak (created, incidentally, by an intrusion of magma 50 million years ago) as Paul Doss engaged in some of what he called "arm-waving geology."

"I want you to look at the Gallatin Mountains," he proclaimed, sweeping his arms through the air. "The reason that they end right there is because they were obliterated by a volcanic explosion."

Doss doesn't look so much like a geology professor as he does a biker who just parked his Harley to play a gig with a bar band on his blues harp. Actually, he is and does all those things. He has a full salt-and-pepper beard. His gray ponytail flows out from under a sweat-stained cap that reads "Yellowstone Association Institute." Tattoos show beneath his short sleeves, and he wears a small gold hoop in his left ear. Between geological exhortations, he smokes a custom molasses- and cherry-flavored blend in a small curved-stem pipe.

*Most visitors, in fact, are sanguinely ignorant that one of the greatest explosions of all time blew Yellowstone to smithereens, or that they are standing three to eight miles above a gigantic reservoir of molten rock.*

Until recently, Doss was Yellowstone's supervisory geologist, but he quickly tired of the bureaucracy and political pressure on the park's scientists. "The uniform just got kind of tight," he explained. So he returned to the Department of Geology and Physics at the University of Southern Indiana. But he continues to teach a three-day course. Doss's thesis is this: Yellowstone is what it is because of the underlying geology. And the biggest geological event in Yellowstone's history was a series of three massive eruptions 2.1 million years, 1.3 million years, and 640,000 years ago.

As Doss likes to exclaim, waving his thick arms: "It's all about the volcano!"

— . —

Of course, the visitors who throng to Yellowstone are unaware of all this. Most, in fact, are sanguinely ignorant that one of the greatest explosions of all time blew Yellowstone to smithereens, or that they are standing three to eight miles above a gigantic reservoir of molten rock, or that the volcano is primed and potentially dangerous. Says Doss, "The vast majority of those 3 million visitors don't realize it's an active volcano."

No, they know Yellowstone simply as one of the most amazing natural areas on Earth.

Yellowstone is a national treasure. It is our first national park. At more than 2.2 million acres, essentially a square measuring more than fifty miles on each side, it is also one of the largest. (In the lower forty-eight, only Death Valley National Park is larger.)

Because Yellowstone is so big and was protected before it was too badly degraded, its ecosystem resembles—and indeed functions—much as it did in the days before Europeans settled the country, clearing forests, plowing land, and exterminating the largest and most troublesome critters. Grizzlies, elk, black bears, wolverines, and lynx still roam the park, just as they did when Lewis and Clark passed nearby more than two centuries ago. The only large mammal exterminated, the gray wolf, was reintroduced in 1995. The park is the only place in the lower forty-eight where bison survived in the wild. The Yellowstone herd is more or less free-roaming. (If too many bison leave the park and can't be chased back, they are shot. So much for freedom to roam.) Driving into Yellowstone from the west along the Madison and Gibbon rivers, I was reminded how beautiful the park is. Indeed, I was salivating over the possibilities of fly-fishing for trout in the numerous runs, riffles, and pools when suddenly I hit a traffic jam, as drivers ahead stopped to gawk at a bull elk lying in the grass thirty yards off the highway, its enormous antlers covered in velvet.

Like all of those other tourists, I came to Yellowstone to see its geologic wonders—the gushing geysers, hot springs, and all the other heat-generated phenomena that were largely responsible for the designation of the first national park back in 1872. Altogether, Yellowstone has more than three hundred geysers, the largest concentration of such features on Earth.

But unlike other visitors, I was keenly aware that the park straddles one of the largest volcanoes ever. And that raised fascinating questions: What are the chances the volcano will reawaken and erupt again in an earth-shaking, caldera-forming explosion? And if it does, what will happen to the park? And to the people living around the park? And to the people living in the western United States?

Or, for that matter, to the rest of us? Would a volcano put a temporary halt to the world's agriculture and cause widespread famine? Could it drive a cycle of extinctions?

But before I got to those questions, I wanted an on-the-ground, arm-waving sort of introduction to the geology of the park. I wanted to understand how an active volcano, one of the largest ever to exist, has shaped the park we see today. And how the notable features of the park—the geysers and hot springs and, more subtly, the meadows and forests—are a manifestation of that volcano.

— . —

At the base of Bunsen Peak, hills and low mountains all around us, Doss explained that Yellowstone's volcanism is caused by an immense hot spot issuing from somewhere deep within the earth. Its point of origin is not entirely clear. It may flow upward from the lower regions of the earth's iron-rich fluid mantle, nearly two thousand miles

*Imagine slowly passing your hand over a candle flame. In that way, the hot spot has burned a path along the bottom of North America.*

below where we stand. Or it may arise from just beneath the crust, a bit more than one hundred miles below. Regardless of its source, this hot plume of rising, molten rock burns into the earth's solid crust from below, creating a huge chamber of molten rock beneath our feet.

This hot spot seems to be stationary relative to the interior of the earth. But the earth's surface is not stationary. The continents and ocean beds are made up of several large plates and many more small plates, floating on the denser mantle below and zigzagging relative to one another. The North American Plate, like a ship of granite on a malleable basaltic sea, has been sailing toward the southwest at a bit more than an inch a year—about the rate your fingernails grow.

When the Yellowstone hot spot first appeared nearly 17 million years ago, it burned through the crust under what is now northwestern Nevada. Since then, the hot spot has stayed where it is, but the North American Plate has moved to the southwest. Imagine slowly passing your hand over a candle flame. In that way, the hot spot has burned a path along the bottom of North America. Since we perceive the continent as stationary, it is easier to think of the hot spot burning a track from Nevada to the northeast.

Periodically, the hot spot melted a tremendous magma chamber in the crust, not far below the surface. Occasionally, the magma worked its way to the surface, leading to a devastating eruption that partially emptied the magma chamber. The ground above the evacuated chamber plunged down like a tremendous piston, leaving a crater called a caldera.

During the past 17 million years, according to one recent study, the hot spot has produced 142 of these caldera-forming eruptions and countless smaller eruptions that filled the calderas, eventually creating a level fifty-mile-wide volcanic plain—what we now recognize as the Snake River Plain—angling northeast through southern Idaho and leading to the doorstep of Yellowstone. The three most recent super explosions occurred in or very near what is now the national park.

As the crust passed over the hot spot, something else happened—the land stretched and bulged upward a third of a mile. A wave of faults and earthquake zones formed along the leading edge of the bulge and off to either side and trailing behind. It is called a tectonic parabola. "This is not unlike the bow wave of a ship plowing through the ocean," explained Doss. "Sometimes it is called the parabola of death."

Most volcanoes aren't caused by Yellowstone-style hot spots. Rather, as in Indonesia, they form near subduction zones, where dense basaltic sea plates grind slowly beneath buoyant granitic continental plates. Nonetheless, hot spots are not uncommon. There are about thirty on Earth, and most lie beneath the sea. Well-known hot spots created Iceland, the Hawaiian Islands, and the Galápagos Islands. The plates that make up the sea floor are made of basalt. Basalt is

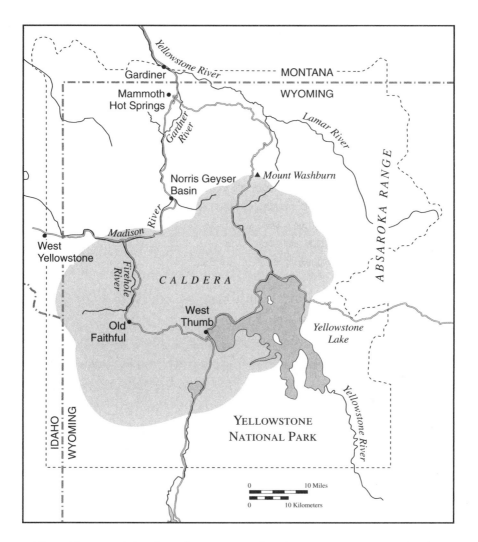

The caldera remaining from the super eruption 640,000 years ago occupies the center of Yellowstone National Park. Lava from subsequent, but much milder, eruptions have largely filled the caldera, creating the Yellowstone Plateau.

black and dense, and when it melts over a hot spot, it oozes and flows like mud. Hence, you can watch as lava flows slowly across the landscape of Hawaii and live to tell the tale.

Yellowstone's is the only major hot spot beneath a continental plate. Continental plates are made mostly of rhyolite. (Rhyolite is volcanic rock made of basically the same stuff as granite, but granite cools slowly underground to form large crystals.) Because rhyolitic magma contains a lot of silica (Yellowstone rhyolites are more than three-quarters silica dioxide), rhyolite does not flow, even when molten. It is viscous, like pine sap. So magma chambers filled with rhyolite don't ooze lava like the volcanoes of Hawaii. Instead, they block the building pressure until the force is uncontainable. Then, watch out.

When the rhyolitic magma does let loose and the pressure decreases, carbon dioxide and water vapor degas from the magma with the explosive force and speed of dynamite. It is that expansion and force of the degassing, rather than the pressure of the rising magma, that actually causes the catastrophic force of the eruption. (A volume of magma, with a water content of 5 percent by weight, expands 670 times as it is raised from the high-pressure depths of the earth to the surface. So a lump of magma the size of an airline carry-on bag would expand to fill your living room with water vapor, magma foam, and magma. The explosive process of expansion would, of course, destroy your house.) Doss said, "If you were to melt that granite, you would be setting the stage for some phenomenal cataclysmic explosions." And that is exactly what the Yellowstone hot spot has done—142 times.

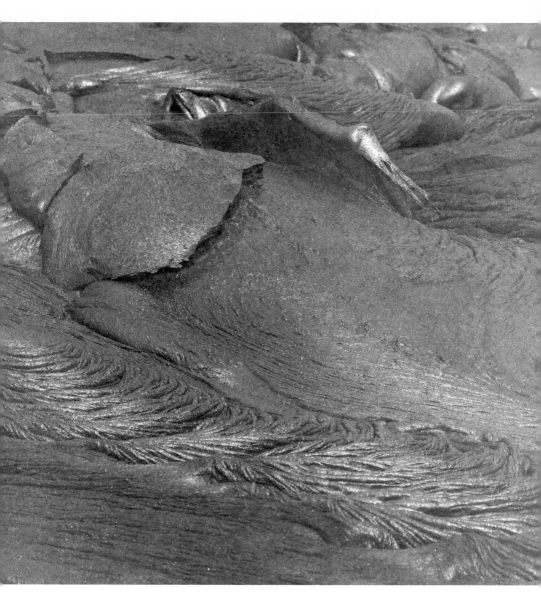

The basaltic lava oozing across the Hawaiian landscape is fueled by a hot spot that issues from deep within the earth. Where a hot spot underlies a rhyolitic continental plate, as in the case of Yellowstone, the lava does not ooze. Instead, the viscous rhyolite triggers some of the biggest explosions in the recent history of the earth. *Bryan Busovicki, Shutterstock*

All evidence suggests Yellowstone is just a stop along the way. The crust continues to travel over the hot spot. In time, Yellowstone will fill with flows of basalt and look like the flat Snake River Plain. People of the distant future will grow potatoes in Yellowstone. They will look for geysers and grizzlies—if grizzlies still survive—in what is now the state capital. All the action will have moved to the northeast. Or, more properly, the crust will have marched to the southwest. Said Doss, "Billings is essentially headed this way!"

— . —

Not all of Yellowstone's topography arose from the cataclysm of its massive caldera explosions or even in the lava flows that continued to fill the caldera for hundreds of thousands of years afterward. There were other geologic forces at work.

We climbed aboard the bus and drove west over the spectacular gorge of the Gardner River onto the Black Tail Plateau, where woods mix with green meadows. We parked, piled out, and hiked up a slope to a large boulder—Frog Rock, named for no good reason, Doss said, except that in the proper light, it looks like the head of a frog. But that's not important, he said. What is important is that this gray chunk of granitic gneiss, roughly the size of an outhouse, doesn't

*People of the distant future will grow potatoes in Yellowstone. They will look for geysers and grizzlies—if grizzlies still survive—in what is now the state capital.*

belong here. It isn't now and never was attached to the geology beneath our feet. It was carried here from the Beartooth Mountains by the conveyor belt of ice known as a glacier.

No continental glaciers reached as far south as Yellowstone. And the Yellowstone Plateau, though high, was not high enough to generate its own glaciers, Doss explained. But the Beartooth Plateau, at more than ten thousand feet, even higher in elevation than Yellowstone, was a factory of glaciers during the most recent Ice Age. Glaciers spilled off the Beartooth and flowed westward down river valleys such as Soda Butte, Hell Roaring, and Slough Creek until ice eventually covered Yellowstone. Ice continued to travel downhill, carrying Frog Rock and other glacial erratics into Yellowstone. As ice piled up, Yellowstone grew cold enough to manufacture its own glaciers. Ice traveled north to Chico Hot Springs, west down the Madison Valley, south to Jackson Hole, east to Pahaska Teepee.

Piles of sand, gravel, and boulders traveled with the ice and accumulated at the ice margins where the glaciers ended. When the ice melted by about 14,000 years ago, glacial debris lay scattered across the ground. In fact, said Doss, we were standing on a deposit of glacial sand and gravel. How could we tell?

Because, he answered (knowing we would not be able to), this is a meadow, covered with grass. Glacial sediments trap water and provide the basis of good soil—soil that can be colonized and built by the deep-rooted grasses and meadow forbs that surround us. Their root systems are so deep and dense that trees can't complete. The grassland, of course, provides graze for bison and elk—just one example of how geology governs biology, Doss said, as the wind buffeted the bunch grasses and flowers.

Contrast this grassland, he said, with the forest we will see inside the volcano caldera. There the soils are broken-down rhyolite, so silicaceous and poor that only drought-tolerant, shallow-rooted lodgepole pine can grow there. There are acres and acres of almost nothing but lodgepole. "The weed tree that grows where nothing else wants to grow," Doss explains. "The ghetto tree—*Pinus monotonous.*"

Throughout the morning, we drove east along the Lamar Valley, a mix of forest, broad meadows, with the rollicking Lamar River—in Doss's words "one of the coolest places on Earth." Doss's perception was colored by several factors. First, the valley contains a mix of interesting geologic features, such as the Precambrian outcrop of granitic gneiss he told us to lay our hands on so we could appreciate the oldest thing in the vicinity, which formed in the unimaginable past, before the hot spot, before even older Yellowstone volcanoes, before even significant life in the earth's seas. A second reason to love the Lamar: It's a terrific place to spot elk, bison, even wolves. Osborne Russell wrote in his *Journal of a Trapper*, "There is something in the wild romantic scenery of this valley which I cannot . . . describe; but the impressions made upon my mind while gazing from a high eminence on the surrounding landscape one evening as the sun was gently gliding behind the western mountain and casting its gigantic shadows across the vale were such as a time can never efface from my memory." And a third reason: Doss had been pulling some really large cutthroat trout from the Lamar River of late.

We crossed a meadow of grass and big sage and began to scale a steep hillside. The hot sun and steep terrain stratified the class by ambition and condition. As we climbed, we found shards of petrified wood—remnants, Doss said, of a redwood forest, much like that of coastal California that once covered these hillsides. As we scrambled higher, we found ever larger pieces, the size and heft of bricks. The rock looked amazingly like wood, and I often had to touch and lift it to convince myself it was not simply a splinter from a modern-day tree. Finally, high on the hill stood a stump, two feet across and four feet tall, looking for all the world like a tree just felled by a lumberjack. Yellowstone's petrified trees were discovered early on. A mountaineer named Moses "Black" Harris told the story of a petrified forest with leaves and branches and even birds frozen in midsong. Mountain man Jim Bridger did him one better: "That's peetrifaction. Come with me to the Yellowstone next summer, and I'll show you peetrified trees a-growing, with peetrified birds on 'em a-singing peetrified songs."

Yellowstone's petrified trees were buried by flows of mud, ash, and sand—a volcanic flow of muck known by its Indonesian name *lahar*. Silica in the sediments replaced organic tissue, creating a lifelike fossil. The landslides were triggered by volcanoes that predate the caldera explosions. About 50 million years ago, long before the hot spot even formed, movements of the earth's plates created an orgy of mountain building and geologic activity that pushed up more than a dozen eruptive centers in Montana and Wyoming, now known as the Absaroka volcanics. Back then, the North American Plate, such as it was, ended not far to the west so that Yellowstone was near the coast. The subduction of the oceanic plate under the continental plate created a range of volcanic mountains, much as the volcanic Cascade Range formed inland of the subduction zone in Seattle's backyard.

Among those was nearby Mount Washburn, a classic stratovolcano. A stratovolcano is what everyone thinks of when they hear the word *volcano*. Lava rises from deep within the earth and gushes from a central point, forming layer-by-layer of lava, cinders, and ash a steep-sided symmetrical cone in the image of Mount Fuji. Other cones of this sort include Mount Shasta in California, Mount Hood in Oregon, Mount Rainier and Mount St. Helens in Washington—all volcanoes associated with a subduction zone, usually where a continental plate overrides an oceanic plate. As Mount Washburn erupted, it triggered the lahars that leveled and buried the Yellowstone redwoods.

— . —

At a highway cut into the flanks of Mount Washburn, Doss showed us the tilt of the layers of lava and ash as they flowed from the crater of the volcano about 50 million years ago. We can sight along them, he said, and look right back at the volcano's peak. Glowing hot pyroclastic flows that had roared down the slope at up to two hundred miles per hour were now frozen.

"Right from the vent of Mount Washburn!" Doss exclaimed. You would expect, he said, that if we could stand on the northern slope of the mountain and see the layers of

Clouds enshroud Mount Fuji, a stratovolcano in Japan. Built layer-by-layer of lava, cinders, and ash issuing from a central point, it rises to 12,388 feet. It last erupted in 1707–08. *Hiroshi Ichikawa, Shutterstock*

rock angling back toward the peak, we could be able to do the same on the southern slope. But we can't, he said. Because the southern half of Mount Washburn is missing.

Back on the bus, we rode to where the south slope of Mount Washburn would have been—over Dunraven Pass, at a bit more than 8,800 feet, to the Washburn Hot Springs Overlook. We scrambled from the bus and crowded at the edge of the viewing area. We were perched on a rim, looking far away to the south. Below us spread a flat plain covered with *Pinus monotonous* (in fact, *Pinus contorta*, and here, the largest continuous tract of lodgepole pine in North America).

"You're on the caldera wall," Doss announced. "Mount Washburn happened to be a stratovolcano that happened to be in the wrong place at the wrong time. . . . This half of the volcano was obliterated. It literally chopped Mount Washburn in half with the development of the Yellowstone caldera wall. So you're sitting on it right now."

From where we stood, we had a long, broad view. The weather was sunny, the air clear. The horizon was crisp, and a long way off. "You're on the caldera wall here," Doss reiterated. Then he began to describe the dimension of what we were looking at. "So the caldera wall kind of heads off to the northeast for a little bit." Yes, we could follow it off to our left. "And you can very well delineate the spine of the Absaroka Mountains. See them out there?" Indeed we could—Avalanche Peak, Pyramid Peak, Pelican Cone, perhaps twenty miles away. Then, Doss said, "If you look south, you'll see two mountain ranges. You'll see the Red Mountains are the mountains that are right over Hayden

Valley. The high peak at the east end is Mount Sheridan. The caldera wall is the base of Mount Sheridan."

This took a while to sink in. I understood we were looking across the center of the Yellowstone caldera, the void left by the evacuation of the most recent super eruption. And I understood that we sat on one edge of the caldera, where the southern half of Mount Washburn was obliterated. And I understood that Mount Sheridan, way out there somewhere, marked the other side of the caldera. But for the longest time I thought he was talking about a ridge in the middle distance. No, Doss said, that's Elephant Back. Mount Sheridan is that thing way out there!

On the horizon! Then suddenly the picture came into focus. He was talking about that mountain over there. Mount Sheridan sat more than thirty-three miles from where we stood. (I figured that out later on a map.) Some 640,000 years ago, a vast lake of magma beneath us—as far across as I was looking right now, way out there by the horizon—blew sky high. And then this entire region—again all the land I could see nearly to the horizon—plunged downward for several hundred yards, with attendant explosions, eruptions, landslides, concussive atmospheric shocks, and explosions of displaced magma, like nuclear detonations.

"Oh my god," I said.

"Oh my god is right," said Doss. "That's a good thing to think. . . . This is the only place in the park, the only place anywhere, where you can even begin to piece together the magnitude of this beast. It's massive."

— . —

Doss calls his course Yellowstone's Geoecosystem. "You don't have a biological system without the geological underpinnings! All this charismatic megafauna crap you hear about," he fulminated one day in the field, even as a red fox sneaked behind him and skulked into the sagebrush, "whether it's mayflies, trout, peregrine falcons, grizzly bears, heat-loving organisms that live in the vents of boiling springs, all life is ordered by the underlying rocks. Geology drives everything."

And the biggest geologic driver, of course, is the Yellowstone hot spot, which has elevated the Yellowstone Plateau and the surrounding mountains some 1,700 feet higher than they otherwise would be, so that now Yellowstone National Park averages 8,000 feet above sea level. This high country makes for a cool climate and catches eastward-blowing precipitation. In the southwest corner of the park, nearly one hundred inches of rain and snow falls each year. Gardiner, in the north, gets only a tenth as much.

The difference between nutrient-poor caldera soils and richer glacial soils and lake sediments influences the abundance of plants and animals. On the more fertile soils in the grassy meadows and mixed-species forests, Doss said, there is greater food for animals ranging from bison to squirrels. And that influences the number of predators, such as wolves and bears. Bears dig up soil looking for red squirrels and their stashes of whitebark pine nuts. Their digging prepares the soils for even more diverse plant life.

Just north of Yellowstone Lake, we stopped at LeHardys Rapids, a long pitch of white water on Yellowstone River. From the boardwalk, we could look down into

*The discontinuity of the Gallatin Range, the elevation of the Yellowstone Plateau, the location of the lodgepole pine and the spawning runs of cutthroat trout—all these products of the Yellowstone hot spot are subtle in comparison to the more than ten thousand spectacular geysers, hot springs, and mud pots for which the park is famous.*

the clear water and see five large native cutthroat trout, each about eighteen inches, finning in a quiet eddy. Once rested, they would assault the final pitch of the rapids.

The Yellowstone hot spot even dictates whether trout can ascend the rapids and gain access to Yellowstone Lake upstream. In the 1920s, Doss said, engineers surveyed the route before building the park road. Fifty years later, they resurveyed and discovered their benchmarks were off. Further investigation showed the land had risen fully two feet. What was going on? Two areas were stretching and lifting. One area was Mallard Lake Dome, just north of Old Faithful, about twenty-five miles to the west. The other was the Sour Creek Dome, next to LeHardys Rapids. These were so-called resurgent domes, located near the old vents of the Yellowstone volcano. As magma and hot water surged underground, the domes rose and fell. As these domes, especially Sour Creek, rose, the stairstep of faults now forming the cascades of LeHardys Rapids rose as well. Since the rapids was effectively the lip, or outlet, of Yellowstone Lake, the movement affected the lake level. Depending on its elevation, the rapids became a barrier to trout and determined whether cutthroats had access to the lake and its spawning streams. The evidence of stream sediments and old lake beaches showed that, once, the river had been as much as fifteen feet lower. At other times, the lake was nearly fifty feet higher. Exclaimed Doss, "Here we are talking about the magma dynamics eight miles below the crust and it's influencing the ecology of migrating cutthroat trout."

— · —

The discontinuity of the Gallatin Range, the elevation of the Yellowstone Plateau, the location of the lodgepole pine and the spawning runs of cutthroat trout—all these products of the Yellowstone hot spot are subtle in comparison to the more than ten thousand spectacular geysers, hot springs, and mud pots for which the park is famous. These features exist because heat flow from the area within the old caldera is seventy times more than the average around the world. In the park's active geyser basins, the heat flow is several hundred times greater.

One morning Doss led us down a trail off the Upper Terrace Drive near Mammoth. We crept through the woods to the edge of Narrow Gauge Terrace. The conifer glade seemed enchanted. Sunlight filtered through steam rising from the bubbling pools rimmed by white travertine. Appearances to the contrary, Doss explained, the water was well below boiling—about 160 degrees Fahrenheit. (Boiling point at this altitude is 199 degrees.) The bubbling is caused by the release of carbon dioxide gas as the charged water reaches the surface—like the fizz of a just-opened bottle of 7UP.

Warm mineral-laden waters issuing from hot springs deposit layers of the soft rock travertine, forming terraces at Yellowstone's Mammoth Hot Springs. *TTphoto, Shutterstock*

These terraced pools are some of the world's best examples of travertine-depositing hot springs. They are the result of all the right ingredients: heat, abundant water, limestone, a plumbing system of fractured rock, and frequent earthquakes to keep the plumbing clear. Rain and melting snow seep into fractures. The water is warmed deep underground by superheated brine warmed by magma even farther below. Hot water dissolves carbon dioxide gas escaping from the magma chamber to form a solution of carbonic acid that dissolves the limestone bedrock, the bed of Paleozoic seas, and transports calcium carbonate to the approximately fifty hot springs at the surface near Mammoth. As pressure is released and carbon dioxide bubbles away, the limestone precipitates out of solution and recrystalizes as the soft rock travertine, the building stone of Rome's Coliseum. Layers build at the rate of nearly a quarter inch a day.

The trickling water streams through bands of color. As long ago as 1889, geologist Walter Harvey Weed guessed that the colors were the result of various species of algae. In 1966, University of Wisconsin professor Thomas Brock discovered *Thermus aquaticus*, a bacterium that inhabits these hot-water features. Since then, scientists have discovered hundreds of species of thermophiles living in water ranging from scalding to boiling. Some of these organisms were photosynthesizing bacteria and cyanobacteria. Others were Archaea, an ancient lineage of single-celled organisms without nuclei long believed to be bacteria but recently put in a kingdom all their own. Some of these organisms were extremophiles in other regards as well, tolerating not only high temperatures, but also high levels of acidity or alkalinity or high concentrations of calcium carbonate. Yellowstone's hydrothermal pools range in pH from Crested Pool at 9.4 to Sulfur Caldron at 1.3, between lemon juice and stomach acid. A scientist from Montana State University recently told an audience at Yellowstone that the range of conditions is so extreme, so toxic, "that if it wasn't in a national park, it would be an EPA Superfund site."

The key to surviving such high temperatures lies in the amino acids of the enzymes of these creatures, which retain their structures despite high temperatures. These heat-tolerant enzymes have come in handy in biotechnology. In 1983, DNA chemist Kary B. Mullis discovered that Professor Brock's bacteria could be used in a polymerase chain reaction to make a billion duplicates of a strand of DNA in a few hours. "The DNA may come from a hospital tissue specimen, from a single human hair, from a drop of dried blood at the scene of a crime, from the tissues of a mummified brain or from a 40,000-year-old wooly mammoth frozen in a glacier," Mullis wrote in *Scientific American.* For his discovery, Mullis won the 1993 Nobel Prize in chemistry.

The bands of colors radiating from a hot spring do indeed represent various organisms. As water flows from the spring, it cools, and critters segregate by temperature. Most hold to a narrow range of conditions, such as those found in a single pool in a single terrace. Scientists from the University of Illinois studying one hot spring at Mammoth detected 221 unique species. Because of their precise temperature requirements, there was little downstream colonization.

Which begs a question: How did they get here if they can live only in one narrow range of conditions? Scientists have discovered that thermophilic bacteria can live in the steam rising from their hot-water habitats. But how can they survive as the steam cools and disperses? "They had to somehow or other be brought here," Doss said. "And the only way to be brought here, you're not in hot water. So how did that happen? We don't know the answer to that."

Yellowstone's thermophiles resemble thousands of species of microbes that live near volcanic hot-water vents in the ocean floor, many of which metabolize hydrogen or sulfur for their sustenance. Scientists suspect that life began with these kinds of organisms in just these conditions. By current reckoning, life falls in three major kingdoms—Bacteria, Archaea, and Eukarya. Bacteria and Archaea comprise only microorganisms. Eukarya comprises everything else, from mushrooms to humans (and some thermophilic fungi, such as *Curvularia protuberata*). The links to a common ancestor, which existed perhaps 4 billion years ago, were all hyperthermophiles, which thrive in water warmer than 176 degrees Fahrenheit. "When you look at the organisms here, you're very close to the root of the tree of life," Doss said.

Likewise, if ever we find life on a planet such as Mars, it will probably be some extremophile. For that reason, University of Houston scientists have examined the travertine structure at Narrow Gauge to try to anticipate what the remnant rocks of extremophiles on the Martian landscape might

*When you look at the organisms here, you're very close to the root of the tree of life.*

look like. "So if we study these guys," Doss said, "and we study the kinds of textures they create in the rocks, the evidence that they leave behind, then we might be better prepared to identify the organism that's on a different planet."

— · —

Norris Geyser Basin was a vision of Dante's inferno, a vaporous plain baked in midday sun with turquoise blue basins and steaming vents. In the pan itself was hardly a shred of vegetation—hardly anything at all but scorched earth. Over time, some of these hot springs and geysers go dormant; others spring to life. Norris is, said Doss, the hottest and therefore most dynamic geyser basin in the park. The temperature in a hole drilled to 1,087 feet below the surface measured 459 degrees Fahrenheit. I breathed in the thick smell of sulfur and thrilled to the throaty roar of the Black Growler Steam Vent.

We filed down a boardwalk and, with Doss as lookout, sneaked off into the woods when we thought no one would notice. As a class we had permission to bushwhack and we had a ranger with us to prove it, but neither Doss nor the ranger wanted to have to argue about it. Leaving the boardwalk was forbidden and dangerous without supervision. Even with supervision. "Almost anyone who has done research in the backcountry on volcanic features has at sometime broken through," said Doss. Including Doss. He escaped major injury, but others haven't been as lucky. Doss had rules for walking. Vegetation generally indicates the footing is sound. Bison tracks do not. "You cannot follow bison tracks," Doss said. "I can show you plenty of places where there's a bison skeleton in a hot spring. They don't know where to walk."

We hiked single file through a scraggly stand of lodge-pole pine. Brittle plants poked from the dust. We reached a ridge overlooking the geyser basin, with weirdly clear ponds. One bubbling pond Doss called Coyote Shit Spring for a scat preserved in silica from the hot spring. "It was just beautiful," Doss said. Unfortunately, someone snatched it. "It was gorgeous," he said.

During the afternoon we walked the backcountry to see some of Norris's off-the-beaten-track features. The combination of hot water and fractured bedrock has pro-duced a terrific diversity of hot springs, geysers, mud pots, and fumaroles. The waters are a riot of weird colors, the result of organisms keyed to various regimes of temperature and acidity. The water is also colored by minerals—gray sinter, yellow sulfur, red and black arsenic compounds, and gray and black iron sulfides.

All these features are driven by the heat of the magma beneath Yellowstone, which in turn is created by the 17-million-year-old hot spot. Critical factors for all hydro-thermal features: a supply of water, fractured bedrock to form a plumbing system between the deep source of heat and the surface, and frequent earthquakes to shake the plumbing occasionally to keep it from plugging with min-eral sinter.

Hot springs are the most common of the hydrothermal features, and the least complicated. Water heated deep below the surface rises through underground fractures and

Old Faithful and Yellowstone's other iconic hydrothermal features are powered by heat rising from the subterranean magma chamber—the same reservoir of molten rock that triggered super eruptions in the past. The volcanic history of the park has far-reaching effects in the Yellowstone ecosystem, determining the habitat of everything from tiny heat-loving bacteria to one-ton bison. *Robynrg, Shutterstock*

emerges at the surface as a gushing spring or quiet pool. Water at the surface cools and sinks back into the ground as hotter water rises to maintain the pool at boiling or near-boiling temperature.

Fumaroles are the hottest features in Yellowstone's geyser basins. Unlike hot springs, fumaroles are located above the water table and don't have much water in their plumbing. Surface water draining into the fractures vaporizes and escapes at the surface as hissing steam.

---

*At the slightest disturbance, sulfur precipitate billowed like the downiest of feathers.*

---

Mud pots are also short of water. Hydrogen-metabolizing thermophiles called *Sulfolobus* convert hydrogen sulfide gas to sulfuric acid. The acid breaks down rock and glacial sediments into clay. The thickness of the mud depends on the water supply, which is often linked to rainfall and snowmelt. Gas blurps through the mud to the surface. Mud pots even have a microscopic ecosystem: A virus has been discovered that parasitizes the *Sulfolobus*.

These hydrothermal food chains link to larger ecosystems. In the afternoon, we hiked onto One Hundred Springs Plain. From the parched earth emerged Realgar Spring, its waters laden with iron, arsenic, and mercury. A cyanobacteria lives in the chemical brew, forming a green-black mat. Ephydrid flies lay their eggs in the microbial mat. As the eggs hatch, the larvae eat the cyanobacteria. The larvae turn to pupae, and the pupae transform to adults, which mass and feed on the microbial mat. Though the flies can withstand temperatures of only 109 degrees Fahrenheit, they insulate themselves in a bubble of air to feed in water that is much hotter. The masses of flies—as many as five hundred adults and one hundred thousand larvae per square yard—feed wolf spiders, which dash out across the hot mats. Other creatures also feed on the flies—dolichopodid flies, dragonflies, wasps, beetles. Killdeer prowl the margins of hot springs such as Realgar to eat insects. They, in turn are eaten by birds of prey, such as peregrine falcons. Said Doss, "It's all about the volcano!"

We visited two more springs that were otherworldly. Sulfur Spring trembled from the inflow of warm water. At the slightest disturbance, sulfur precipitate billowed like the downiest of feathers.

Finally, we stepped on stones across a warm stream and hiked to what looked like nothing more than a cauldron of bubbling black beans, forty feet across. Technically, Cinder Pool is an "acid-sulfate-chloride boiling spring." It is covered with a layer of black cinders. They are made of sulfur and pyrite—"some goofy weird product of some reaction with the sulfur deep within this," said Doss. "Take this for what it's worth, but I'm not aware of any feature in the world that's comparable."

— · —

Geysers are Yellowstone's showboats. Like normal hot springs, they have an ample supply of hot water. Silica dissolved in hot water becomes layered in the rock fractures to form piping able to withstand high pressure. Unlike hot springs, geysers have some kind of obstruction in the plumbing—contorted, narrow passages like bad sinuses, or a deposit of sinter plugging the pipes. As steam bubbles upward, the volume is too great to pass easily through the constrictions. So as the water is constrained, its pressure may rise to more than 2,400 pounds per square inch. Finally something gives. The pressure overcomes the resistance of the constriction and the weight of the overlying water. As the pressure drops, the superheated water flashes to steam, sending the water at the geyser's surface spraying into the air. In seconds, or minutes, as the volume of gushing water depletes the water in the plumbing or when the steam is able to bubble to the surface without expelling water, the geyser subsides. Depending on the supply of hot water and the constrictions in the passageways, geysers may be regular. Or they may be as unpredictable as Steamboat Geyser in Norris Geyser Basin, which erupts at intervals from four days to fifty years. But when it does go off, watch out. It is the tallest geyser in the world, throwing water more than three hundred feet high. Even regular geysers can change unpredictably, sometimes from the effects of distant earthquakes.

The most famously regular of Yellowstone's geysers is, of course, Old Faithful, named by the Washburn expedition in 1870 for its predictable eruptions. According to one survey, Old Faithful scored highest as the attraction people most wanted to see. (Bears ranked second.) Our bus raced toward the Upper Geyser Basin, the world's largest concentration of geysers, to catch its next performance. But by the time we reached the parking lot, we saw the plume of steam and water soaring over the rooftop of Old Faithful Inn. "Well," said Doss, "there you go. You see it."

Old Faithful is a cone geyser, spouting water from a single nozzlelike vent. It erupts more often than any other major geyser, though Steamboat is larger, Grand is taller, and others are more regular. Each eruption lasts ninety seconds to five minutes, jetting to a height of 108 to 180 feet and expelling 3,700 to 8,400 gallons through a vent measuring two feet by five feet. In full eruption, water races at supersonic speeds through an underground slot measuring four inches by several feet long. Under high pressure, the water reaches temperatures of 244 degrees Fahrenheit. Steam gets even hotter.

While you wouldn't set a watch by their predictions, park rangers can calculate the next eruption based on the duration of the previous eruption. Shorter eruptions eject less water, and so the plumbing takes less time to recharge and build pressure. Intervals between shows range from 53 to 205 minutes. The average 92 minutes between eruptions is longer than it used to be. The interval increased after the 1959 Hebgen Lake earthquake, increased further after the 1983 Borah Peak earthquake, increased yet again after a small earthquake in 1998, and lengthened once again after a subsequent swarm of quakes. The quakes shake the geyser's plumbing, affecting the water supply or constrictions.

Old Faithful wasn't expected to erupt again for nearly two hours, so Doss and I walked the boardwalk to check out the other geysers. More than half of the geysers in the world are found in Yellowstone, and most of those are located here, along three miles of the Firehole River.

"Every condition that is necessary to create a geyser or a system of geysers is perfectly in place here," Doss was saying. As we strolled the boardwalk, we noticed a bison track within a yard of the edge of Crested Pool, a pool of almost constantly boiling water forty-two feet deep. It boils more violently than others and may build a head during a rolling boil of more than eight feet.

Sawmill Geyser was named for the violent, circular swirling of water in its crater, which looks like a spinning blade. It was erupting as we passed. Surprisingly, the steam from the geyser is like a hot breath but the spray of water is icy. A park naturalist reminded us, both of us bespectacled, that the mineralized water would accrete silica film on our glasses.

---

*Every condition that is necessary to create a geyser or a system of geysers is perfectly in place here.*

---

Alongside these geysers runs the Firehole River—at first glance a gorgeous trout stream, riffling and sparkling. I had thoughts of hooking a trout and dangling it to poach in a nearby geyser. But Doss disabused me of the idea. At this time of year, in mid July, especially when the stream flow is low, this stretch of the Firehole runs too warm to hold cold-water fish, which seek refuge in small icy tributaries.

On the boardwalks and pathways throughout the basin are "geyser gazers," volunteers who have "adopted" various geysers. They monitor the commencement, duration, and conclusion of each eruption of their favored geysers, and they often transmit their raw data to colleagues via two-way radios mounted cop-style on their shoulders.

A crowd of people—geyser gazers and less devoted sorts—had gathered around Grand Geyser. It was still awhile before Old Faithful was scheduled to erupt, so we hung out at Grand. An ancillary geyser began to spout hot water. "Surf's up!" someone yelled. Then the main geyser began to spew. The plume grew to some two hundred feet amid constant cheers. For perhaps ten minutes, the water burst into the air, forming sheets of spray and a slowly pulsating fountain.

As we followed the boardwalk back toward Old Faithful, Doss told me a story from Lee Whittlesey's book *Death in Yellowstone: Accidents and Foolhardiness in the First National Park*. A dog jumped out the window of a parked truck and leapt into the first pond it came to. Unfortunately, this being Yellowstone, the pool was a hot spring. The friend of the dog's owner got to the scene first and jumped in after. "He actually made it to the dog," Doss said. But that's about all he could do. He had to be fished from the pool. His last words: "That was pretty stupid, wasn't it?"

— · —

Whittlesey's book chronicles the multitude of mishaps that can befall Yellowstone visitors. Many visitors assume that being in a national park affords them some safety. Still, they are mauled by bears, gored by bison, frozen in blizzards, and burned in fires. Yellowstone is a dangerous place. That it is also a national park makes little difference.

That said, the accidents of greatest interest to Doss (and to me) for the purposes of his class were those related to the ongoing effect of the Yellowstone hot spot. It has been 70,000 years since Yellowstone's volcano has actually gushed lava to the surface. But as the geysers and mud pots attest, the volcano is still active and poses a present danger.

On August 17, 1959, just minutes before midnight, two earthquakes—6.3 magnitude and 7.5 magnitude—struck within seconds of each other near Hebgen Lake, just west of the park. A mountainside slumped, sliding more than a hundred miles per hour, wiping out part of the Rock Creek campground and generating winds of hurricane velocity. The landslide plunged into the Madison River, sending thirty-foot waves upstream and down. Twenty-eight people died, including nineteen buried in the landslide, never to be recovered. Huge waves in Hebgen Lake wiped out several homes on the shore. The ground on one side of a fault shifted up to twenty-two feet relative to the land on the other side. Some geysers in Yellowstone began erupting for the first time, or more furiously than before.

Aside from earthquakes, the most likely violent events are so-called hydrothermal explosions. These are not volcanic, as such. Superheated water and steam build pressure beneath the surface, like a super geyser. But the pressure never finds a vent. Instead, it builds until it clears away the obstruction, the bedrock ductwork, and nearly everything else in the vicinity. In 1989, Porkchop Geyser in Norris Basin exploded, heaving rocks more than two hundred feet.

Past hydrothermal explosions have been huge (though nothing in comparison to Yellowstone's far-distant volcano).

Castle Geyser in Yellowstone ejects water and steam every nine to eleven hours. Yellowstone has not gushed lava for 70,000 years, but its famous geysers attest that the volcano is still active and poses a present danger. The park contains more than three hundred geysers, the greatest concentration of such features on earth. *Bartosz Wardzinski, Shutterstock*

An explosion 13,000 years ago excavated Yellowstone Lake's Mary Bay, well over a mile across and nearly a mile wide and 50 feet deep. Not surprisingly, the hottest spot on the lakebed, 252 degrees Fahrenheit, was discovered in Mary Bay.

Perhaps the dynamics of a hydrothermal explosion are easiest to imagine at Indian Pond. It's smaller than Mary Bay by far. But it's an archetypal example of what happens when trapped steam and superheated water have nowhere to go. The pond is nearly circular, about a quarter mile across, and is surrounded by a berm of earth, the so-called ejecta rim of expelled materials.

"One good thing about ejecta rims: the crap that is blown out of the crater and then is deposited radially around the structure, because if it lands on top of organic material and then kills it, if you sample that organic material, you can have it radiocarbon dated," Doss said. Indian Pond blew out a bit over three thousand years ago. Since, the rim has been covered by grass and forbs, and grazed by countless bison.

We followed a trail along the rim of the pond. A bison at water's edge grazed—unaware or unconcerned. But we were keenly aware and kept watch for any aggressive behavior. Of all the park's hazards, from exploding steam vents to grizzly bears, among the most frequent and least expected are aggressive bison. One of our class members was also a park naturalist. "Watch their tail," she said. "If it goes up, it means one of two things: charge or discharge."

— · —

For our last stop, our bus dropped into the valley of the Firehole in the Lower Geyser Basin. We pulled off at a gravel trail, and Doss led us on a short walk along the banks of the river. Soon we stared into a blurping mud pot. It lay at the base of a ridge that, Doss explained, was the ejecta rim of yet another hydrothermal explosion.

Doss was introduced to this place a couple of weeks after starting as the park's supervisory geologist. "I had the same reaction as you: This is a beautiful place. This is a gorgeous place. And in fact over the ejecta rim boundary over there is an extensive mud pot field. It's wonderful." There are other features not many tourists get to see, including springs with delicate draperies of sinter. But Doss didn't bring us here to show us all that. He had a story to tell.

Park employees and others in the know would often sneak to this place to lounge in the hot springs at the edge of the Firehole. One of their favorites was a spring whose scalding waters overflow and commingle with the cool current of the river. Bathers have constructed a stacked-rock wall that controls the mixing and creates a pool that's just the right temperature for "hot potting."

"That's what it's referred to—like sitting in a hot tub," said Doss. "So this little bathtub right here is a pretty famous place for hot potting."

One summer afternoon in 2000, several park summer workers were hot potting along the Firehole at this very place. Afternoon turned to evening. Three of the hot-potters, a twenty-year-old woman and two eighteen-year-old men, scrambled downstream along the river to get back to their car. Unfortunately, none had a flashlight. The moon had yet to rise, and the night was black. Suddenly, the other members of the group heard screams and shouting.

"They got to here," Doss said. He led our group along the route the three young people would have taken, to the edge of another spring, a deep, clear pool more than ten feet across, at the edge of the river.

"Apparently, the three locked arm in arm," Doss said. As a survivor explained later, they looked down at the crusted soil and saw a patch of light—where they were standing—then a band of dark, and then a band of light. They believed the dark was a small stream flowing into the Firehole. So they joined arms to jump across.

But it was not a stream. And the light patch beyond— that was the reflective surface of the spring. There they landed, side by side. The shortest of them, the young woman, on the right, landed in the very deepest part of the spring.

"So they locked arms and they jumped over the stream," Doss said. "And they jumped right into the pot. Two guys—they fell, but they got out by themselves. The girl went completely under and actually ended up swimming to the other side. She tried to get out over here. . . . I mean there was a lot of screaming at this point. Some people had come over to help. She was dragged out. And pretty quickly she was actually dragged down into the river in an attempt to cool her. She died fifteen hours later. It was probably a good thing they put her in the river because that put her in shock. That was cold enough, that it probably just put her in shock. She was in there for forty, fifty minutes. So it probably gave her hypothermia. Which was

Hot springs, like geysers, are formed when the magma chamber boils water that flows to the surface. Geysers form when the hot water is contained by an obstruction in the geological plumbing until building pressure jets the water skyward. *Fernando Rodrigues, Shutterstock*

probably in the end a very good thing. . . . It was suggested the girl was probably in at least seven seconds. And that's a very long time."

As Doss told the story, I stood at the water's edge and looked into the clear water. Scalding temperatures make water no more or no less clear. The spring measures 178 degrees Fahrenheit. I tried to imagine the temperature—like the hottest tap water scalding your hands, but perhaps thirty degrees hotter. Water that hot over your entire body. For seven long seconds. Slowly count to seven while hot water swirls with the flailing of your arms and legs.

The young woman was burned over her entire body; the young men each over 90 percent. The men, with help, walked out to the road. The woman was carried. All were flown to the Salt Lake City Burn Center. The woman died the next day. "The guys amazingly have survived," said Doss, who went to the scene the next day to help with the investigation. He remembered, he said, the footprints of the victims with the dried blood and splattered fluid from their wounds. "I can't walk this trail without seeing those images."

— . —

The next day, we found a stretch of river conducive to happier memories. We clambered down a steep bank of the Gardner River just downstream from Mammoth. The river valley divided into braided channels, shallow and bouldery for the most part, with endless yards of unimpeded space in which to unfurl a fly line.

Doss concentrated on the smaller sloughs, while I pummeled pocket water in the main channel. From time to time, I'd glance at Doss. We were catching a few small rainbow trout, though not as many as you'd think by looking at the water, which seemed excellent.

Like most everything else in Yellowstone, the Gardner River is a product of its geology. Because of the rise of the Yellowstone Plateau, pushed up over eons by the Yellowstone hot spot, the stream rises on a high slope near the northern edge of the caldera rim and runs northward, gathering up Straight, Winter, Indian, and Panther creeks before flowing through the Mammoth Hot Spring basin. There the cool creek waters mix with the hot, alkaline

---

*I tried to imagine the temperature—like the hottest tap water scalding your hands, but perhaps thirty degrees hotter. Water that hot over your entire body. For seven long seconds. Slowly count to seven while hot water swirls with the flailing of your arms and legs.*

---

waters of Hot River. During the last Ice Age, a glacier flowed off the plateau down the valley of the Gardner. And most of the river's course skitters over and cuts through the Lava Creek Tuff deposited by the most recent caldera-forming explosion, as well as the basalts laid down by later eruptions. So the waters of the Gardner are a product of all of this—glacial till, calcareous spring waters, minerals of volcanic rock. As a sum, they are the Gardner—a rollicking trout stream, producing invertebrates that trout eat and possessing the pools and runs where trout, including native cutthroats, seek cover and spawning habitat.

I marched upstream through islands of outwash rubble and cobble, covered with forbs and willow—perfect setting for a grizzly. The day was getting on; I had to leave for home. But up ahead was a long, deep pool where the river tumbled over a five-foot ledge and then ran fast against a cliff. Right away I lost a trout about a foot long. Then I landed the best fish of the day, a fifteen-inch rainbow. I tied on a large black Wooly Bugger with plenty of weight and drifted it through the pool, near the cliff. The line stopped—stopped cold as if I had snagged a rock. But it began to move, and then it set to tearing up the pool, running here and there. It was a bull of a trout. I put as much pressure as I dared on the light tippet. I saw the fish briefly—perhaps eighteen inches, broad across the back. Then suddenly the hook came out. The line went slack. As I watched a big shadow vanish into the deep pool, my knees shook.

It's all about the volcano.

# Chapter 3

# Natural Wonders

Humans have long recognized the wondrous features of Yellowstone. But it took a while to understand what really caused them.

Yellowstone's volcanic eruptions had gone stone cold by the time prehistoric hunters found their way to the plateau with the retreat of Ice Age glaciers some 14,000 years ago. Still, it must have been an exciting time, with melt waters fueling steaming geysers and boiling springs, and thundering hydrothermal explosions blasting out Indian Pond and other craters, heaving tons of rock and earth to the sky.

These Paleo-Indians hunted with spears tipped with big points chipped from nearby supplies of obsidian (a volcanic glass). They killed the last of the Pleistocene megafauna—the mammoths and giant bison—as well as the elk and bears that would survive into the modern era. Forests of lodgepole pine and meadows of prairie grasses were reclaiming the barren plain left by the glaciers. In the thousands of years since, people have camped around Yellowstone Lake and hunted in the park. Oral traditions place the ancestors of the Salish in Yellowstone. And the Shoshone say their people originated here. About three thousand years ago came the invention of the bow and arrow, and with it, hunters more efficiently killed smaller animals. The hunters built sheep traps and bison corrals.

Many historic tribes lived or hunted in the park, or traveled across the plateau. They bathed and cooked in the hot springs. The Bannock used their namesake trail through the northern part of the park to cross from their home in the Snake River Plain, where bison had disappeared, to hunt in eastern-lying plains. Other tribes— Blackfeet, Crow, Flathead, Cayuse, Coeur d'Alene, Nez Perce, Umatilla—also crossed the plateau. A group of Shoshone lived in the highlands, even as other tribes procured horses and hunted bison on the plains. They ate mostly bighorn sheep and became known as Sheepeaters.

Clearly, these early visitors were savvy to the geysers and hot springs. Writes historian Joseph Owen Weixelman: "Hundreds of years before the first Euro-Americans gazed on the Firehole valley, many American Indians went to the geyser basins to pray, meditate, and bathe." Many of the hundreds of Indian campsites in the park are located near hot springs and geysers. Shoshone and Bannock collected pigment in the springs. Shoshone soaked the horns of bighorn sheep in the hot water to form their bows. The many names for Yellowstone reflected their knowledge of the area. The Shoshone name meant "water keeps on coming out." The Blackfeet called it "many smokes." Yellow Wolf, one of Chief Joseph's scouts in the Nez Perce War, said the "hot smoking springs and the high-shooting water were nothing new" to his tribe.

---

*Hundreds of years before the first Euro-Americans gazed on the Firehole valley, many American Indians went to the geyser basins to pray, meditate, and bathe.*

---

It was through reports from local Indians that white explorers first heard of the wonders of Yellowstone—the "boiling waters volcanoes," in the words of one trapper. Father Francis Xavier Kuppens, a Belgian Jesuit, wandered the Missouri basin with Piegan (Blackfoot) Indians during the mid 1860s. "It was while leading this nomad life that I first heard of the Yellowstone," Kuppens wrote. "Many an evening in the tent of Baptiste Champagne or Chief Big Lake the conversation, what little there was of it, turned on the beauties of that wonderful spot."

While Indians knew of the jetting water and thermal pools, it's less clear what they thought caused them. Printed accounts have been filtered through the stereotypes and ignorance of early white travelers. Lewis and Clark, who passed north of the park and never saw its geysers, said the Indians they met reported that "very frequently there is a loud noise heard like thunder which makes the earth tremble—they state they seldom go there because their children cannot sleep at night for this noise and conceive it possessed of spirits who are averse that men should be near them."

Missionary Father Pierre-Jean De Smet said, "The hunters and Indians speak of it with a superstitious fear, and consider it the abode of evil spirits, that is to say, a kind of hell. Indians seldom approach it without offering some sacrifice, or at least without presenting the calumet of peace to the turbulent spirits, that they may be propitious. They declare that the subterranean noises proceed from the forging of warlike weapons: each eruption of earth is, in their eyes, the result of a combat between the infernal spirits, and becomes a monument of a new victory or calamity."

As Weixelman rebuts, the Indians had no metallurgy, and so no notion of "forging" weapons: "This anecdotal myth seems to derive more from the pagan Greeks than with American Indians. . . . Most native peoples revered the land of Yellowstone and many treated it as sacred in their cosmology. While a sense of fear may have been linked with the geysers and hot springs, the belief that this was the predominant emotion or indicated a primitive intellect is mistaken. Instead, it is more accurate to say that American Indians understood the area to be linked to the powers of their Creator, powers that were difficult to understand and could be dangerous."

Yellowstone took its name from its namesake river. The river, in turn, was named by the Minnetaree Indians for the sandstone bluffs near present-day Billings (not the yellow rock of the Grand Canyon in the park, as some people believe). The Minnetaree name was directly translated by the French as *roche jaunes* and by English cartographer David Thompson as "yellow stone."

The first white man known to find the area was John Colter, who split off from Lewis and Clark and made a grueling winter trek in 1807–08 through Yellowstone. Later, he described "hot spring brimstone" from at least one hydrothermal basin. Clark continued to add notations to his map of the area as new reports came from fur trappers.

Year by year, the hydrothermal features became better known to explorers and cartographers. At about the time Colter roamed the area, Governor James Wilkinson of the Louisiana Territory claimed to have organized an exploratory party by "Perogue" up the Missouri and Yellowstone rivers. Little information of their discoveries survives, except for a map of the upper Yellowstone that local Indians drew on a buffalo hide. The hide was presented to President Thomas Jefferson with a letter that noted, "a Volcano is distinctly described on Yellow Stone River."

After the War of 1812, more fur traders reached Yellowstone. A party from the North West Company traveled within sight of the Tetons in 1818. Their scribe, Alexander Ross, reported that "boiling fountains having different degrees of temperature were very numerous; one or two were so very hot as to boil meat." Trapper Daniel Potts noted "hot and boiling springs, some of water and others of most beautiful fine clay . . . [that] throws its particles to the immense height of from twenty to thirty feet in height." Potts' account in *The Philadelphia Gazette and Daily Advertiser* in 1827 was the first description of Yellowstone in the media.

Beaver trapper Joseph L. Meek traveled with a party of trappers along the Yellowstone River in about 1829. A band of Blackfoot scattered the trappers. Nineteen-year-old Meek fled across the Yellowstone with a mule, blanket, and gun. He wandered southward into what is today the park, where "behold! the whole country beyond was smoking with the vapor from boiling springs, and burning with gasses, issuing from small craters, each of which was emitting a sharp whistling sound," according to his recollection to Francis Fuller Victor in *The River of the West*. The hot springs stirred thoughts "about hell and the day of doom."

Warren Angus Ferris, a clerk of the American Fur Company, added substantially to the knowledge of Yellowstone's hydrothermal features. He traveled to Yellowstone in 1834 more out of curiosity than business, and for that he has been dubbed Yellowstone's first tourist. "From the surface of a rocky plain or table, burst forth columns of water of various dimensions, projected high in the air, accompanied by loud explosions, and sulphurous vapors, which were highly disagreeable to the smell. . . . The largest of these wonderful fountains, projects a column of boiling water several feet in diameter, to the height of more than one hundred and fifty feet. . . . These explosions and discharges occur at intervals of about two hours. After having witnessed three of them, I ventured near enough to put my hand into the water of its basin, but withdrew it instantly, for the heat of the water in this immense chauldron [*sic*], was altogether too great for my comfort; and the agitation of the water, the disagreeable effluvium continually exuding, and the hollow unearthly rumbling under the rock on which I stood, so ill accorded with my notions of personal safety, that I retreated back precipitately, to a respectful distance. The Indians, who were with me, were quite appalled, and could not by any means be induced to approach them. They seemed astonished at my presumption, in advancing up to the large one, and

*The hot springs were not only spectacular, but convenient for cooking, where the "kettle is always ready and boiling."*

when I safely returned, congratulated me on my 'narrow escape.' They believed them to be supernatural, and supposed them to be the production of the Evil Spirit."

The following year Osborne Russell visited Yellowstone with mountain man Jim Bridger and his trappers. He wrote what was till then the most detailed description of the hydrothermal features: "Near where we encamped were several hot springs which boil perpetually. Near these was an opening in the ground about 8 inches in diameter from which steam issues continually with a noise similar to that made by the steam issuing from a safety valve of an engine and can be heard 5 or 6 Mls. distant. I should think the steam issued with sufficient force to work an engine of 30 horse power." On a subsequent trip in 1839, they came upon the West Thumb geyser basin, including "about 50 springs of boiling hot water," including at least one active geyser. The hot springs were not only spectacular, but convenient for cooking, noted Russell, where the "kettle is always ready and boiling." Russell also described a "boiling lake" a hundred yards across. "The steam which arose from it was of three distinct Colors from the west side for one third of the diameter it was white, in the middle it was pale red, and the remaining third on the east light sky blue. Whether it was something peculiar in the state of the atmosphere the day being cloudy or whether it was some Chemical properties contained in the water which produced this phenomenon.

I am unable to say and shall leave the explanation to some scientific tourist who may have the Curiosity to visit this place at some future period. . . . What a field of speculation this presents for chemist and geologist."

— . —

Trappers and prospectors continued to explore Yellowstone. In 1863, about forty prospectors accompanied Walter Washington deLacy to investigate the headwaters of the Snake River. A civil engineer, deLacy drew up a map used by the Territory of Montana to draw out the borders of its first counties. He contributed further specific knowledge of the geysers and hot spring.

Still, detailed scientific knowledge of Yellowstone was spotty. But the dribble of reports and tantalizing details spurred the organization of more comprehensive surveys. These came in a burst, from 1869 to 1871.

In 1869, David E. Folsom, Charles W. Cook, and William Peterson—friends and business associates in the goldfields of Montana—followed the divide between the Gallatin and Yellowstone rivers into Yellowstone. Their observations were the most detailed to date, and drew the strongest connection between the thermal wonders and previous volcanic activity in the area. "Half a mile down the river, and near the foot of the bluff, was a chalky-looking bank, from which steam and smoke were rising; and on repairing to the spot, we found a vast number of hot sulphur springs," Cook and Folsom wrote in *Western Monthly Magazine*. "The steam was issuing from every crevice and

Henry D. Washburn, surveyor general of the Montana Territory, led the 1870 expedition to what would become Yellowstone National Park. Washburn realized volcanic activity formed Yellowstone and correctly predicted the area would someday draw throngs of tourists. *William Henry Jackson, National Park Service*

hole in the rocks; and, being highly impregnated with sulphur, it threw off sulphuretted hydrogen, making a stench that was very unpleasant. All the crevices were lined with beautiful crystals of sulphur, as delicate as frost-work. At some former period, not far distant, there must have been a volcanic eruption here. Much of the scoria and ashes which were then thrown out has been carried off by the river, but enough still remains to form a bar, seventy-five or a hundred feet in depth. Smoke was still issuing from the rocks in one place, from which a considerable amount of lava had been discharged within a few days or weeks at farthest. While we were standing by, several gallons of a black liquid ran down and hardened upon the rocks; we broke some of this off and brought it away, and it proved to be sulphur, pure enough to burn readily when ignited."

The men also had the prescience to anticipate the popularity of Yellowstone Lake: "It is a scene of transcendent beauty, which has been viewed by but few white men; and we felt glad to have looked upon it before its primeval solitude should be broken by the crowds of pleasure-seekers which at no distant day will throng its shores."

Importantly, members of the Cook-Folsom expedition discussed the then-novel idea of reserving part of the area as a national park. Folsom went to work for General Henry D. Washburn, surveyor general of the Montana Territory. Folsom not only fed Washburn important details of Yellowstone; he also suggested the area should be set aside for public use.

In 1870, Washburn himself set out for Yellowstone. With him were Montana politician and businessman Nathaniel P. Langford (later first Yellowstone superintendent) and

Cornelius Hedges (attorney, publisher, and would-be gold miner). They were accompanied by a small attachment of soldiers from Fort Ellis led by Lieutenant Gustavus C. Doane. The Washburn-Doane expedition visited Yellowstone Lake and the geysers along the Firehole River, naming many, including Giant, Giantess, Castle, and Old Faithful. They measured and analyzed, bringing back details of the area and helping to refine the maps of the region. Washburn wrote of the experience in "The Yellowstone Expedition" in the *Helena Daily Herald*. And Doane reported one of the most thorough accounts of Yellowstone to date. Wrote U.S. geologist Ferdinand V. Hayden, "for graphic description and thrilling interest it has not been surpassed by any official report made to our government since the times of Lewis and Clark."

*It is probably the greatest laboratory that nature furnishes on the surface of the globe.*

Washburn clearly associated Yellowstone's geology and hot springs with volcanic activity. Traveling upstream along the Yellowstone River from the Grand Canyon, they "found three hills, or rather mountains, thrown up by volcanic agency, and consisting of scoria and a large admixture of brimstone. These hills are several hundred feet high, and evidently are now resting over what was once the crater of a volcano."

Both Washburn and Doane figured the area would become a popular spot with tourists. Yellowstone Lake, wrote Washburn, was "a beautiful sheet of water, with numerous islands and bays, and will in time be a great summer resort, for its various inlets, surrounded by the finest mountain scenery, cannot fail to be very popular to the seeker of pleasure, while its high elevation and numerous medicinal springs will attract the invalid." Doane predicted an even greater purpose: "As a country for sight seers, it is without parallel. As a field for scientific research it promises great results, in the branches of Geology, Mineralogy, Botany, Zoology, and Ornithology. It is probably the greatest laboratory that nature furnishes on the surface of the globe."

Langford lectured on the exploration first in Helena and Virginia City, and then in Washington, D.C., New York, and Philadelphia. These presentations, especially the Washington lecture, stirred interest in further exploration of Yellowstone among those who attended, including prominent politicians and Ferdinand Hayden.

In 1871, the year after the Washburn expedition, Hayden led the first full-fledged government expedition to Yellowstone. His task was to give "attention to the geological, mineralogical, zoological, botanical, and agricultural resources of the country." Hayden had no shortage of help in the endeavor. His party eventually included an assistant, agricultural statistician and entomologist, topographer, assistant topographer, meteorologist, botanist, assistant botanist (who would later become a congressman), mineralogist and medical doctor, another physician, zoologist, several general assistants, and Army engineer John W. Barlow.

Accordingly, Hayden brought back the most detailed scientific information to date. He recognized Yellowstone

U.S. geologist Ferdinand Hayden led the first comprehensive federal expedition
to the park in 1871 and returned East with the most detailed scientific findings
to date. He recognized the long history of volcanism in Yellowstone and even
recognized the structure that scientists would later discover was a monstrous
caldera. *William Henry Jackson, National Park Service*

as the focus of a huge volcanic assemblage and noted the frequent small earthquakes. He even hinted at the existence of a sprawling caldera: "From the summit of Mount Washburn, a bird's-eye view of the entire basin may be obtained, with the mountains surrounding it on every side without any apparent break in the rim. . . . It is probable that during the Pliocene period the entire country drained by the sources of the Yellowstone and the Columbia was the scene of as great volcanic activity as that of any portion of the globe. It might be called one vast crater, made up of thousands of smaller volcanic vents and fissures out of which the fluid interior of the earth, fragments of rock, and volcanic dust were poured in unlimited quantities. . . . Indeed, the hot springs and geysers of this region, at the present time, are nothing more than the closing stages of that wonderful period of volcanic action that began in Tertiary times."

Importantly, Hayden also brought along photographer William Henry Jackson and artist Thomas Moran. Hayden worked diligently during the year after his return to promote the creation of Yellowstone National Park. The designation came so quickly that Moran had barely had a chance to paint. But his dreamy paintings and Jackson's photography would have their effect later—to publicize this newly protected geological wonderland to the American public.

— . —

In 1865, as Belgian priest Francis Kuppens traveled with Piegan companions, he was stormbound at St. Peter's Mission near the mouth of the Sun River with several other

*Summit of Jupiter Terraces*, taken in 1871, and other dramatic photographs by William Henry Jackson were instrumental in imprinting the newly discovered geological wonderland in the American mind. *National Park Service*

travelers, including acting Territorial Governor Thomas Francis Meagher. As they passed the time, Kuppens described the strange wonders of a place called Yellowstone. A Canadian also taking refuge piped up with even more details. Meagher was clearly impressed and said if all that were true, the government ought to set the area aside as a national park.

Meagher's remark was the earliest recorded in print that Yellowstone should be protected as a national park. It was a novel idea, since till that time, there were no national parks. Not in this nation, or in others. The closest model was Yosemite in California, granted to the state by the federal government in 1864. (Yosemite was later designated a national park in 1890.)

Yellowstone fulfilled all the requirements of a great national park; as the first, it set an incredible precedent. A vast space of wild country, it fulfilled Thoreau's requirements when he asked, "Why should not we, who have renounced the King's authority, have our national preserves, where no villages need be destroyed, in which the bear and panther, and some of the hunter race, may still exist, and not be 'civilized off the face of the earth' . . . not for idle sport or food, but for inspiration and our own true recreation?"

But more than its vastness, its pristine unsettled condition, or the fact that it was inhabited by grizzlies, bison, and other disappearing creatures was this fact—Yellowstone contained otherworldly geysers and hot springs. During the late 1800s, hardly anyone was concerned about the area's importance to wildlife. Or that it comprised an intact natural ecosystem (a concept unrealized then). That it would provide a place for scientific study was interesting but not vital. What was important was that Yellowstone contained "wonders" that excited the public appetite for the *sublime*.

The sublime was a romantic appreciation of wilderness. It originated, tellingly, where there was no wilderness—in the cities of Europe and the eastern United States. The sublime was the power of nature to inspire awe. With the blossoming of the Romantic period, the attention of artists turned from the beauty of the village and pasture and focused on untrodden wilderness, where the lighting was dramatic, the trees towering, the rocks craggy, and the weather frightful. Humans, if they appeared in the work of art, were often natives. If European, they were dwarfed and humbled by their surroundings.

---

*The sublime was a romantic appreciation of wilderness. It originated, tellingly, where there was no wilderness—in the cities of Europe and the eastern United States.*

---

"The sublime dispelled the notion that beauty in nature was seen only in the comfortable, fruitful, and well-ordered," writes Roderick Nash in *Wilderness and the American Mind*. "Romanticism," writes Nash, ". . . implies an enthusiasm for the strange, remote, solitary, and mysterious." And what could be more remote and mysterious than a high plateau surrounded by mountains in the unexplored West, where the earth rumbled, springs bubbled, and geysers erupted on schedule? Yellowstone, it seemed, was a manifestation of both heaven and hell. What could better exemplify the sublime? Such were the attitudes and the power that underlay Thomas Moran's sublime interpretations of Yellowstone. Only later did politicians and others begin to think of the value of wild country independent of its sublime volcano-driven wonders of nature—in the words of Senator William B. Bate of Tennessee, "primeval nature, simple and pure."

The creation of a national park was not without controversy. Then, as now, some politicians doubted the need to protect land of so little immediate economic value. They saw no sense in legislation that might inhibit the construction of roads into the park. Nonetheless, the theory of a public park held in common to prevent development, and the fascination with Yellowstone's natural wonders fused as a singular idea. The designation proceeded with astounding speed. President Ulysses S. Grant signed the law creating the park on March 1, 1872, a scant six months after Hayden's survey. (Interestingly, the designation never gave the park a name. That came later, apparently in correspondence between Superintendent Langford and the secretary of the interior.) Once the park was made, the *New York Times* crowed: "Perhaps, no scenery in the world surpasses for sublimity that of the Yellowstone Valley; and certainly no region anywhere is so rich, in the same space, in wonderful natural curiosities."

— · —

*Tower Falls and Sulphur Mountain* and other paintings of Yellowstone by Thomas Moran applied the sublime aesthetic to the new park's already dramatic landscape. *National Park Service*

Yellowstone's explorers had scratched the surface. They realized the park's "curiosities" had some association with ancient volcanoes. They also suspected that earthquakes were somehow linked to volcanoes. "I have no doubt," wrote Hayden after his 1871 exploration, "that if this part of the country should ever be settled and careful observations made, it will be found that earthquake shocks are of very common occurrence."

Now geologists began to dig deeper. The park's second superintendent, Philetus W. Norris, emphasized the importance of scientific research in the park. His annual report for 1880 included sixty-five pages describing the "true origin" of various geysers and hot springs.

In 1883, members of the U.S. Geological Survey began the first study of Yellowstone's rhyolitic plateau. The work was difficult; the country nearly inaccessible. But they soon realized the plateau was confoundingly complex. Artist-turned-geologist William Henry Holmes, who accompanied Hayden to Yellowstone and accepted the task of surveying the new park's geology, viewed Yellowstone's volcanics as a related series of events.

Working in the late 1800s, American geologist Arnold Hague recognized that "great flows of lava" and intrusions of magma, including a tremendous outpouring of rhyolite, formed the Yellowstone plateau. Hague also suspected that the volcanic activity drove the park's hydrothermal fireworks. "That the energy of the steam and thermal waters dates well back into the period of volcanic action, there is in my opinion very little reason to doubt. . . . All our observations point in one direction and lead to the theory that the cause of the high temperatures of these waters must be found in the heated waters below, and that the origin of the heat is in some way associated with the source of the volcanic energy."

Nonetheless, the scale of the volcanism eluded Hague and so did its source. He thought perhaps Mount Sheridan and Mount Washburn. And he did not associate Yellowstone's volcanism with the long track of calderas across the Snake River basin. While volcanoes wracked the landscape, Yellowstone is "not to be compared in size and grandeur with the volcanoes of California and the Cascade Range," Hague wrote. "Indeed the region may be considered long since extinct."

Later scientists tried to better understand Yellowstone's thermal features. During 1929–30, Arthur Day and E. T. Allen of the Geophysical Laboratory of the Carnegie Institution of Washington bored holes to depths of more than four hundred feet to record temperatures (up to four hundred degrees Fahrenheit) below Upper Geyser and Norris basins. In 1935, Day and Allen published the most complete account of Yellowstone's thermal features in *Hot Springs of the Yellowstone National Park*.

As recently as about 1960, scientists realized the rhyolite of the Yellowstone plateau was the result of an incredible explosion, but they still did not understand its source. Like Hague, they continued to look to conventional stratovolcanoes, such as Mount Washburn.

Until this time, geologists had been handicapped by the "old geology." They may have understood that Yellowstone was the result of a series of volcanoes. But they lacked the knowledge of deep hot spots, or why hot spots might arise, or that hot spots might appear to move steadily and incrementally beneath the crust, or—most important—that the crust itself was moving steadily over the mantle of the earth. They would not begin to more fully comprehend Yellowstone until a revolution had turned geology on its head.

# CHAPTER 4

# EVOLVING GEOLOGY

Solid rock. Rock solid. "The house fell not: for it was founded upon a rock."

"Laws are sand, customs are rock."

The Rock of Our Salvation. "The Lord is my rock."

"The earth abideth forever."

So much of geology occurs outside our frame of reference—so gradually, so slowly, that to believe that rock is less than solid, that the mountains are movable, is counterintuitive.

So it was that not only the public at large, but geologists too believed the continents were unmoving. They might push up mountains or change shape as higher seas flooded their coasts. They might be buried and graded by glaciers. They might be covered by magma.

But they stayed where they had formed billions of years ago.

In the era of old geology, when most everyone believed in static continents, geologists struggled to answer several questions: How were mountains created? Why were marine fossils found on mountaintops? Why did bands of similar geology and fossils seem to run from continent to continent, even though the landmasses were separated by thousands of miles of water? Wasn't it queer, as Francis Bacon had noted four centuries ago, that the maps of the New World showed that coastlines of North America and Greenland, and South America and Africa, fit together like pieces of a puzzle?

To begin to answer some of these questions, a few mavericks suggested that if only we could speed up time and watch from a vantage point far above the earth, the continents would scatter across the oceans like fallen leaves on a windblown pond.

One of the first persons to suggest that continents might be in motion was Austrian geologist Eduard Suess. To explain the similarity of fossils across the boundaries of distant continents, Suess proposed in the late 1800s that an ancestral supercontinent he called Gondwanaland split apart and was separated by oceans. And the fossil beds split apart with it.

American geologist Frank Bursley Taylor took up this controversial idea anew. At the annual meeting of the Geological Society of America in 1908, Taylor reiterated the astounding suggestion that the continents had once been joined but then drifted apart. Note how with only a bit of adjustment, the continents would nestle together—Greenland against the coast of North America, and South America into the west coast of Africa.

Then, in an article two years later for the *GSA Bulletin*, Taylor presented what may be the first detailed discussion of continental movement. Not only do the outlines of the continents match up, Taylor wrote, but so also do the Tertiary mountain ranges. And, as Eduard Suess had noted before him, so do fossils in South America, Africa, and India. Then, remarkably, Taylor took his hypothesis a step further, identifying the Mid-Atlantic Ridge as one of the "most remarkable and suggestive objects on the globe." This line of mountains running north-south through the ocean like the seam of a baseball "marks the original place of the great fracture" from which "the continents on opposite sides . . . crept away." Hardly anyone believed it, of course, though Taylor continued to promulgate his idea of "horizontal sliding of continental crust-sheets" for the next twenty years. It was an outstandingly prescient thought, almost completely ignored.

But the idea attracted at least one person. He was German meteorologist and Greenland explorer Alfred Wegener. In 1912, after an expedition in Greenland in which several members of his four-man party narrowly escaped death, Wegener returned to the University of Marburg. "One day a man visited me whose fine features and penetrating blue-gray eyes I was unable to forget," the great German geologist Hans Cloos later recalled. "He spun out an extremely strange train of thought about the structure of the Earth and asked me whether I would be willing to help him with geological facts and concepts."

Wegener elaborated on Taylor's ideas. He pointed out that the Appalachians of eastern North America aligned with the Scottish Highlands, and that the rock layers of the South African Karoo matched the Brazilian Santa Catarina system. Wegener also noted that fossils were found in bands that reached across continents now separated by thousands of miles of open ocean. Other fossils appear in present-day climates entirely unsuited to them—tropical ferns and cycads in the Norwegian arctic island of Spitsbergen, for example, and Carboniferous-era glaciation in South America, Africa, India, and Australia. He suggested the land had in fact moved to a colder latitude and the changing configuration of landmasses had influenced local climates. In his book *The Origin of Continents and Oceans*, published in 1915, Wegener proposed that for hundreds of millions of years the continents had existed as one supercontinent he

---

*The theory ignores the source of sediment, the nature of rock, and character of most mountain ranges. Other than that, it was brilliant.*

---

dubbed Pangea ("all earth") and then moved apart, a process he called "continental drift." (He elaborated on his hypothesis in several later editions.) "It is just as if we were to refit the torn pieces of a newspaper by matching their edges and then check whether the lines of print ran smoothly across," wrote Wegener. "If they do, there is nothing left but to conclude that the pieces were in fact joined in this way." Wegener pointed out that the continents are made largely of rhyolite, and are less dense than the volcanic basalt that makes up the deep-sea floor. Wegener proposed that the continents floated somewhat like icebergs.

Geologists had been struggling to explain the creation of mountain ranges. Suess himself had argued that the earth was cooling and contracting and that its crust wrinkled on the shrinking earth. American geologist James Dwight Dana suggested the boundaries between continents and oceans absorbed most of the stress of the shrinking earth, and the crust wrinkled along the continental boundaries. James Hall, first president of the Geological Society of America, proposed a hypothesis of the Appalachian Mountains—that as basins, or geosynclines, filled with sediments, they would somehow collapse into folded bedrock and then elevate as mountains. (As one commentator wrote, the theory ignores the source of sediment, the nature of rock, and character of most mountain ranges. Other than that, it was brilliant.) To account for the continuity of fossil deposits on opposite sides of oceans, geologists suggested various land bridges had existed to carry organisms across the seas.

Wegener began by demolishing the theory that large land bridges had once connected the continents and had since sunk into the sea as part of a general cooling and contraction of the Earth. There were few traces of any such bridges. Wegener argued that if mountains were built by a shrinking earth, they would be distributed randomly and evenly around the globe instead of being piled up in ridges, usually along the edges of continents. "The forces which displace continents are the same as those which produce great fold-mountain ranges," Wegener wrote. "Continental drift, faults and compressions, earthquakes, volcanicity, [ocean] transgression cycles and [apparent] polar wandering are undoubtedly connected on a grand scale."

Some geologists applauded Wegener's idea. South African Alexander du Toit found it a plausible explanation for the similarity in fossils and geology between Africa and South America. (Two decades later, he wrote *Our Wandering Continents* to compile evidence of Wegener's theory, but most everyone in North America ignored it.) Swiss scientist Émile Argand thought collisions between continents provided the force needed to raise the Alps. "It placed an easily comprehensible, tremendously exciting structure of ideas upon a solid foundation," wrote Hans Cloos. "It released the continents from the Earth's core and transformed them into icebergs of gneiss on a sea of basalt. It let them float and drift, break apart and converge. Where they broke away, cracks, rifts, trenches remain; where they collided, ranges of folded mountains appear."

But most reaction was hostile. Rollin T. Chamberlin of the University of Chicago called Wegener's hypothesis "of the footloose type, in that it takes considerable liberty with our globe, and is less bound by restrictions or tied down by awkward, ugly facts than most of its rival theories." Prominent American geologist Thomas Chamberlin argued that "if we are to believe [this] hypothesis, we must forget everything we have learned in the last 70 years and start all over again."

Wegener died of an apparent heart attack in Greenland in 1930 (his frozen body found the following year, fully dressed, in his sleeping bag, atop a reindeer skin). At the time, most geologists continued to believe that shared fossils came about because life forms had traveled phantom land bridges exposed by falling sea levels. Despite the shortcomings of these accepted theories, there remained two big obstacles to continental drift: What force moved the continents? (Wegener had suggested the centrifugal force of the earth's rotation and tidal pull, both of which he himself realized were inadequate.) And how could they plow through the solid rock underlying them? Besides, Wegener was a meteorologist. Who was he to attack the literal and figurative foundations of geology?

But evidence that seemed consistent with Wegener's theory slowly mounted. Detailed mapping of the Alps and Appalachians indicated the mountain systems would spread for hundreds of miles if they somehow flattened out. In other words, there wasn't enough wrinkled crust to form them. Detailed measurements of gravity in the early twentieth century suggested that continental crust was less dense than the layers of rock beneath them.

Finally, physicists discovered that heat from radioactivity was enough to keep the earth from cooling. Irish geologist John Joly argued that this radiation-generated heat could melt deep layers of rock. English geologist Arthur Holmes wrote *Principles of Physical Geology* at Durham University during World War II, when a dearth of students (most were off fighting) left him with plenty of time. In it, Holmes argued that radioactivity deep within the earth formed convection currents in the earth's mantle that mobilized the less-dense continents. Holmes also suggested that the inner earth did not need to be liquid to allow convection currents and the movement of continental crust, but only plastic in nature. Holmes acknowledged his ideas were "purely speculative" and would "have no scientific value until they acquire support from independent evidence."

Opposition to the idea of continental drift persisted, especially in North America. But evidence continued to mount. Work in mapping magnetism and gravity in the deep ocean began to explain the mechanics of continental drift and win support for Wegener's ideas.

Beginning in the 1920s, Dutch geophysicist Felix Vening Meinesz, of Delft Technical University, pursued his curiosity about gravitational anomalies in the world's oceans. The task of measuring minute variations in gravitational force would seem especially difficult in a ship, but Meinesz devised a gravimeter consisting of a pair of identical pendulums swinging in opposing cycles. Navigating the oceans off the coast of Indonesia in the Dutch submarines *Her Majesty K II* and *Her Majesty K XIII*, Meinesz measured low-gravity belts that corresponded with the 24,440-foot-deep Java Trench.

At the same time, American scientist William Bowie was working with the U.S. Navy to measure gravity in the oceans. Bowie invited Meinesz to the United States to conduct his work in the deep trenches of the Caribbean and the Gulf of Mexico. Among those who joined the expeditions in the ship *Barracuda* were Americans Harry Hess and Maurice Ewing, and later, Brit Edward Bullard. Their work confirmed Meinesz's earlier findings: Gravity was weaker where the ocean was deep. Hess and Meinesz speculated that downward currents in the mantle and downward deformation of the crust might cause the "negative" gravity. Experiments, including a laboratory model of a crust of paraffin over a sea of oil and convection currents created by two rotating cylinders, showed that such a process was at least possible. Hess seized upon the findings and summarized them in a paper that suggested the convection currents could maintain a downward buckle in the crust if they moved at velocities of one-half to four inches per year. Later estimates for the movement of continental plates fell comfortably within this range.

*The idea that these ridges were like a crack in an egg, occasionally leaking the inner goop of planet Earth, led Hess and Robert Dietz to the theory of sea floor spreading.*

The importance of submarine warfare during World War II compelled greater study of the ocean floor. The military was interested in ways of finding enemy subs and hiding their own. Hess, a member of the Naval Reserve, was called to duty and made captain of the assault transport *USS Cape Johnson*. He was assigned, among other things, to operate a fathometer to measure depth to map the ocean floor. The work continued into the postwar years, a golden age of oceanic research. As Stanford professor Naomi Oreskes writes: "More was learned about the oceans during these 25 years than in the entire previous history of science." This mapping showed long lines of mountains and canyons in the oceans, including the gigantic Mid-Atlantic Ridge that snaked down the Atlantic between the Old World and the New. Seismologists determined the ridges were volcanically active, and geophysicists discovered that the floor of the ocean comprises relatively young rocks, compared with the continents, which are made up of both recent and ancient rock. The idea that these ridges were like a crack in an egg, occasionally leaking the inner goop of planet Earth, led Hess and Robert Dietz to the theory of sea floor spreading.

Maps of magnetism in the ocean floor provided additional insights into the movement of continents and the phenomena of sea floor spreading. Scientists knew that tiny grains of iron would align with the magnetic field and then harden in place as magma cooled, providing a permanent record of their orientation. In 1906, French physicist Bernard Brunhes discovered that this paleomagnetism showed the earth's magnetic field had reversed polarity in the past. Two decades later, geophysicist Motonari Matuyama made a similar discovery as he studied volcanic rocks in Japan: Iron particles in recent lava flows aligned with the current magnetic field, but rocks older than ten thousand years were aligned in the opposite direction. In the early 1950s, Jan Hospers found evidence in Iceland that basaltic lava flows were alternately normal and reversed in their alignment.

In the 1950s, English researchers Patrick Blackett of Imperial College and Keith Runcorn and Edward Irving of Cambridge University examined the magnetic alignment of volcanic rocks of different ages from Australia, India, North America, and Europe. They realized that the rocks had not remained stationary with respect to the earth's poles. Either the poles had wandered, or the continents had. By comparing the magnetic records between continents, the researchers realized the continents had moved in relation to the poles and also with respect to one another. The finding was a huge endorsement of the theory of continental drift.

Wrote Hess, "The general picture on paleomagnetism is sufficiently compelling that it is more reasonable to accept than to disregard it."

Traditionally, geologists gauged ages of rocks in relative terms: Newer rocks lay atop older rocks. Collections of similar fossil assemblages suggested similar ages. Layers of volcanic ash over broad areas provided a common reference point. And so on. The radiometric uranium-lead dating method, developed in the early 1900s, gave geologists a way to absolutely date rocks based on the slow decay of uranium. But given the element's very long half-life, this dating method worked only for very old rocks. During the 1950s, scientists at the University of California, Berkeley, developed the potassium-argon dating technique, which could date rocks only a few hundred thousand years old. By combining this dating with paleomagnetic studies, scientists constructed a timescale of magnetic field reversals.

---

*Mason soon noticed a stunning pattern scroll out of his equipment . . . a series of zebra stripes parallel to the coast, the changes from black to white and back representing magnetic field reversals.*

---

At this same time, English geophysicist Ronald Mason, aware of secret U.S. government mapping of magnetic anomalies in the Pacific, asked if he could conduct his own work. Towing an ASQ-3A fluxgate magnetometer behind the U.S. Coast and Geodetic Survey ship *Pioneer*, Mason soon noticed a stunning pattern scroll out of his equipment. As the ship sailed along the Pacific coast between the Mexican border and the Queen Charlotte Islands, his equipment recorded a series of zebra stripes parallel to the coast, the changes from black to white and back representing magnetic field reversals.

Mason's zebra map, published in 1961, went largely without comment in the scientific world. But then Canadian geophysicist Lawrence Morley read Robert Dietz's theory of sea floor spread, arguing that upward convection at oceanic ridges spewed basaltic magma, forming new sea floor in both directions, away from the ridge. Continuing eruptions push cooling basalt farther and farther from the ridge—two mirror-image conveyor belts of basalt emanating from the ridge and heading in opposite directions across the ocean floor. Dietz's theory suggested the continents didn't have to plow through the mantle but rode atop a moving layer of basalt. It also explained why the sea floor was young—because it was continuously being created. When the sea floor reached a subduction zone of downward currents, it was recycled into the earth. As a result, sea floor dated back no more than about 175 million years; in contrast, continental rocks were often billions of years old.

Morley had an *aha!* moment and immediately wrote and submitted a paper to *Nature*. "If one accepts, in principle, the concept of mantle convection currents rising under the ocean ridges, traveling horizontally under the ocean floor and sinking at ocean troughs, one cannot escape the argument that the upwelling rock under the ocean ridge . . . must become magnetized in the direction of the earth's field prevailing at the time. If this portion of

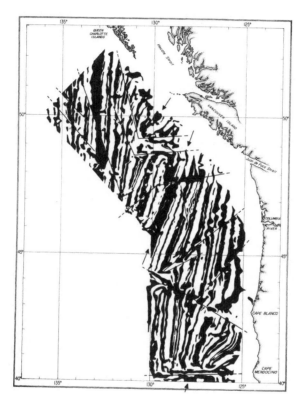

Ronald Mason's zebra-stripe map of the sea bottom along the West Coast proved to be important evidence that new sea floor was continually created and that plates of Earth's crust moved over the mantle. *Ronald Mason and Arthur Raff, Geological Society of America*

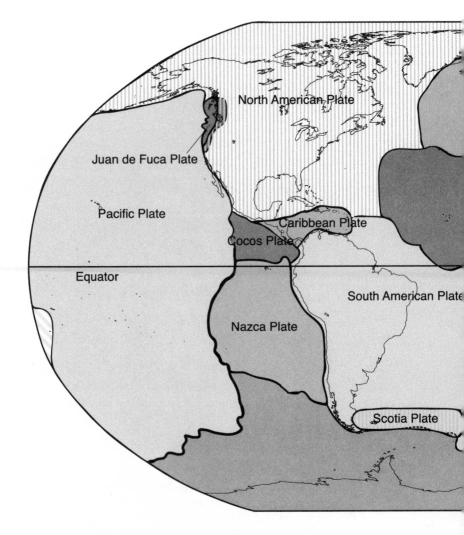

Earth's crust comprises many large plates, propelled in various directions at various speeds by convection currents in the dense but malleable mantle. Collisions between plates produce mountains, volcanoes, and earthquakes. *Adapted from U.S. Geological Survey*

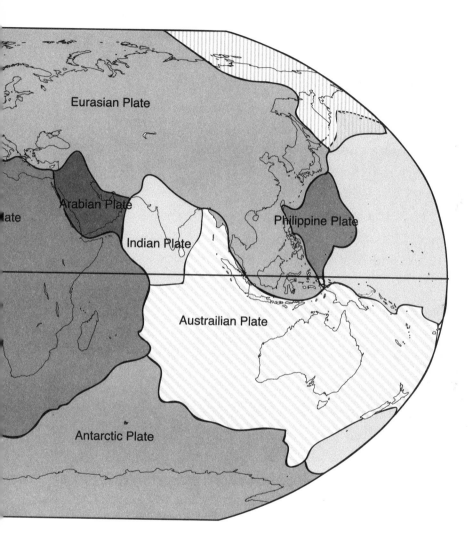

the rock moves upward and then horizontally to make room for new upwelling material and if, in the meantime, the earth's field has reversed, and the same process continues, it stands to reason that a linear magnetic anomaly pattern of the type observed would result." That is, zebra stripes. "It was like finding the key piece to an enormous jigsaw puzzle that made everything fit together," Morley later wrote. Unfortunately, *Nature* did not see the brilliance and rejected the article for lack of space. Neither did the *Journal of Geophysical Research*. Too bad, because Morley had gotten it exactly right. Soon after, *Nature* published a paper by Cambridge geophysicists Frederick Vine and Drummond Matthews that made the same argument.

---

*The earth, instead of appearing as an inert statue, is a living, mobile thing.*

---

Additional work by Vine and Canadian geologist J. Tuzo Wilson showed that the zebra stripes on either side of the Juan de Fuca Ridge were symmetrical to the ridgeline, demonstrating the sea floor spread evenly in both directions. Doubts about the movement of the continents were quickly dissolving. By 1967–68, several articles in the *Journal of Geophysical Research* presented a unified theory, known by the name *plate tectonics*, to explain mountain building, sea floor spreading, and the formation of continents by the movement of more than a dozen large plates over the globe. Wilson eloquently explained, "The earth, instead of appearing as an inert statue, is a living, mobile thing."

By now, the earth looked like this to scientists: At the center of the earth is an inner core of rock 7,000 to 13,000 degrees Fahrenheit that has scarcely cooled since the earth was formed. Despite this heat, the pressure on this inner core is so great that it is solid. This solid inner core is surrounded by a molten outer core. Then, at a depth of about 1,700 miles, begins the lower mantle and then the upper mantle. The mantle is heavy with iron, a sort of über basalt. Within the mantle is a large zone known as the asthenosphere, where, because of heat from radioactive decay, the mantle is plastic and moves in slow convection currents. (Imagine a global lava lamp.) The currents rise to within four to twenty miles beneath the surface, travel sideways, and then sink down again.

From where these currents arise, and how deep they plunge, is not something scientists agree on—perhaps only four hundred miles beneath the surface, perhaps as deep as the core-mantle boundary. On the very surface of the earth, in proportion no thicker than the skin on an apple, is the crust, which ranges to twenty-five miles deep. The crust, together with the uppermost layer of the mantle, is rigid; together they are known as the lithosphere. The lithosphere comprises about a dozen large plates and perhaps twenty smaller ones. With a few exceptions, these plates are either sea floor plates, made of dense basalt, or continental plates, made mostly of less-dense rhyolite. Continental plates are thicker than oceanic ones. Continental plates correspond in an approximate way with the continents themselves. The continental plates ride along on mantle convection currents traveling horizontally beneath the surface.

Upwelling convection currents through the mantle power basaltic eruptions beneath the sea, primarily along the oceanic ridges. This basalt spreads outward as it cools, creating new sea floor. As the creeping sea floor reaches a trench, sinking convection currents pull it back into the earth. This force of cooling currents sinking back into the earth drives the movement of tectonic plates, according to Seiya Uyeda of Tokai University in Japan.

If the sea floor continues to the edge of a continental plate, the dense oceanic plate slides under the thicker but less-dense continental plate and heads back into the deep earth, creating a subduction zone. It is along the world's subduction zones where volcanoes and earthquakes are most common. The Ring of Fire around the Pacific coast is a series of subduction zones, where oceanic plates slide under the continental plates of North America, South America, and Asia.

Because the oceanic plates subduct beneath continental ones, the continents largely keep their shapes. For that reason also, continents just keep getting older. The stable core, or craton, of continents—such as the Canadian Shield or the ancient gneisses of Greenland—date back more than 2.5 billion years, well along toward the formation of the earth 4.5 billion years ago. Collisions between two continental plates push up mountain ranges. When the Indian Plate jammed into the Asian Plate about 50 million years ago, the force raised the folds of mountains we know as the Himalayas.

As theory, plate tectonics gave a revolutionary, plausible, and inclusive explanation of questions that had dogged geologists for a century. Plate tectonics was to geology what Darwin's theory of evolution was to biology, or Einstein's theory of relativity was to physics. It was no less earth-shaking, literally and figuratively, than that.

# Chapter 5

# The Yellowstone Puzzle

When a young geologist named Robert L. Christiansen came to work for the Denver office of the U.S. Geological Survey in 1961, he was assigned to the Nevada testing ground, where his job was to identify rock formations that would best contain the blast and radioactivity of underground nuclear explosions. He discovered that relatively young volcanic rocks did the trick best of all. On the basis of this sort of insight into volcanic formations, Christiansen was sent to Yellowstone in 1965 to study the park's geology.

Christiansen was part of a small team of scientists tapped to conduct the first comprehensive geologic study of Yellowstone since Arnold Hague had researched the park's geology, described in a series of reports published between 1888 and 1912. Geologists were each assigned a topic, and Christiansen and Richard Blank were given the job of examining and describing Yellowstone's young volcanic rocks.

Much of Yellowstone's volcanism occurred about 50 million years ago, with the formation of the Absaroka volcanics, including classic stratovolcanoes such as Mount Washburn. In the mid 1960s, plate tectonics remained a controversial but increasingly well-known theory, and more and more geologists were coming to accept the idea that chunks of the earth's crust sailed around on the mantle rock and that the Absaroka Range had evolved as the North American Plate plowed into the Farallon Plate, triggering a spasm of mountain building and volcanism in the western United States. But those were old volcanoes. Clearly, there were younger volcanic rocks in Yellowstone, evidence of more recent volcanism. What were these rocks, where did they come from, what caused them, how did they evolve, and how were they related to the older rocks? These were the sorts of questions Christiansen and Blank set out to answer. Part of their research was funded by the National Aeronautics and Space Administration. NASA wanted to test its methods of "remote sensing" from space, that is, its imaging of various wavelengths of radiation, such as infrared, from space to identify geology and topography on the ground. And Yellowstone was a good place to "ground truth" its methods. If NASA's satellites could read and discriminate among recent volcanic rocks in Yellowstone, they could be counted on to do the same for the barren rocks of the moon, Mars, and beyond.

When Christiansen and Blank began their work, major questions about Yellowstone's volcanism remained to be resolved. Hayden's work had been preliminary but prescient, calling Yellowstone's plateau "one vast crater." Hague had added to our knowledge of the region's geology immensely. In the 1890s, Joseph Iddings distinguished between the massive rhyolite of Yellowstone's plateau and the other, earlier volcanic rock. Nonetheless, the distinction was not clear even in the minds of geologists. Some continued to blend the two into a single prolonged event. R. A. Daly described the Yellowstone plateau as the "foundered crust of a roofless batholith," accurate as far as it went, but without the recognition of the terrific explosion that caused it.

*The stage was set for Christiansen and Blank to assemble the remaining pieces of the puzzle and tell a coherent volcanic narrative of Yellowstone.*

In 1961, just as Christiansen began his work in Nevada, Francis Boyd explained that Yellowstone's rhyolitic tuffs, one of the most abundant formations in the park, were formed as pyroclastic flows—the outpourings and avalanches of red-hot ash, cinder, and magma from repeated eruptions that had settled in vast sheets and piles while still hot and rewelded into solid rock. These massive tuffs were fractured by "block faulting," which created a basin surrounded by scarps up to three thousand feet high. Then the basin was covered by repeated flows of rhyolitic and some basaltic magma. By mapping their distribution and estimating their thickness, Boyd concluded that these formations—some six hundred cubic miles of welded tuff and other volcanic rocks—made Yellowstone one of the world's great volcanic areas.

Boyd's paper was "an astounding piece of work," Christiansen said recently, with forty years of hindsight. Boyd knew there was a thunderous eruption. He didn't grasp that there was more than one. Said Christiansen, "He came very close to understanding some of these things."

So the stage was set for Christiansen and Blank to assemble the remaining pieces of the puzzle and tell a coherent volcanic narrative of Yellowstone. They went to work. For several summers, they drove, hiked, climbed, and bushwhacked. They needed to survey much of the park, but little of it was served by roads or even hiking trails. Much of the terrain was steep. Even the level plateau was covered by a thicket of lodgepole pine. Downed trees lay heaped and crossed like pick-up sticks. Christiansen and Blank, and an ever-changing team of assistants, looked for exposed rocks in cliffs, road cuts, and eroded creek valleys. They hammered at outcrops and collected samples. They tried to determine which outcrops were continuous with other outcrops, what constituted contiguous flows, which flow lay on top of other flows. They were confounded by the great variety of rock appearances, even within a single lava flow. Some were glassy, others crystalline, depending on rate of cooling. Some were welded more densely than others. These variations were signs that distinguished the order in which flows formed, and how much time had passed between events (sometimes just a matter of hours or days, by appearances). Moreover, later events—such as large explosions of magma—buried earlier lava flows beneath hundreds of feet of rock, obliterating any opportunity to examine the evidence. They sent samples off to Denver, which had one of the best laboratories available for the then-new technique of potassium-argon dating, to determine the ages of each distinct formation. And they developed an inventory of the volcanic geology, published as reports and big colorful maps showing the location and extent of each volcanic formation in Yellowstone and, where relevant, the surrounding area, especially to the south and west onto the Snake River Plain. Christiansen and Blank worked together until 1970; Christiansen continued the fieldwork until 1981.

From the maps describing the area of flows, and by estimating the thickness at various points of the flow, Christiansen was able to calculate the volumes of material belched out by volcanoes. The results were staggering, nearly unknown in the history of the earth.

Christiansen dubbed the oldest and largest of the major volcanic flows the Huckleberry Ridge Tuff. It alone represented a greater explosive volcanic eruption than any known. About 2.1 million years old, it buried an area of 6,000 square miles "in a period of time so short that no erosion and no appreciable cooling of earlier parts of the deposit occurred before completion of the eruption, *certainly within a few hours or days*." The absence of large crystals in the rock (which would indicate slow cooling), or signs that erosion had occurred between flows, suggested to Christiansen one thing: that the Huckleberry resulted all at once in a cataclysmic explosion. He estimated the total volume of magma ejected during the eruption on the order of 600 cubic miles—imagine a cube 8 miles to a side, standing tall enough to block the flight of a jet airliner, simply blown to vapor, gas, and red hot ash and landing in a sheet 500 feet deep, covering an area larger than Connecticut.

The "eruption of such an enormous column of magma . . . in such a short time resulted in collapse of the magma-chamber roof to form a caldera somewhere between 75 and 95 km [nearly 60 miles] long," he later wrote. The caldera stretched from nearly the eastern edge of the park way westward to include Island Park.

But, Christiansen's and Blank's maps declared, Yellowstone was just getting started. A second explosion 1.3 million years ago formed what the geologists called the Mesa Falls Tuff. This explosion appeared to be centered just west of Yellowstone, in Island Park. Though much smaller than the first, it was nonetheless monstrous in comparison with any historic eruption. It ejected 67 cubic miles of magma. (Imagine a cube 4 miles on a side.) The eruption left a collapsed caldera 10 miles across.

Finally, Yellowstone let loose with a third explosion that buried an area of 2,900 square miles. It was bigger than the previous explosion, though not at large as the first, ejecting 240 cubic miles (a cube of molten lava more than 6 miles on a side) "again, probably within a few hours or days." And "like the first two volcanic cycles, [it] produced windblown ash that accumulated in ponds and hollows on the Great Plains of Nebraska and Kansas to depths of as much as nine meters." Once the smoke and ash cleared and the rock cooled, the welded lava formed the Lava Creek Tuff. The collapse of the partially emptied magma chamber created a caldera more than 30 miles across and 50 miles long. As large as that was, it nestled easily within the dimensions of the earliest caldera. In places, the tuff stood as deep as 1,500 feet. This most recent caldera is visible

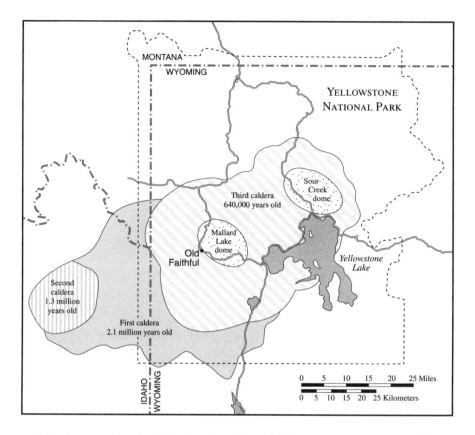

Laborious examination of volcanic tuff revealed three major eruptions of the magma chamber now located beneath Yellowstone National Park. The caldera from the largest eruption 2.1 million years ago nearly contains smaller eruptions 1.3 million and 640,000 years ago. *Adapted from Robert B. Smith and Lee J. Siegel,* Windows into the Earth: The Geologic Story of Yellowstone and Grand Teton National Parks today from vantage points such as Dunraven Pass. Most of the lakes we see today were formed by the backup of streams in valleys blocked by lava. Many waterfalls (such as Lewis Falls, Gibbon Falls, Virginia Cascade) formed as streams rushed over the steep edges of lava flows.

---

*The three caldera-forming eruptions, as brief and explosive as they were, were each part of a grand cycle lasting hundreds of thousand of years.*

---

Christiansen and Blank confirmed that the three caldera-forming eruptions were more recent than most anyone had realized. They all occurred during the periods of glacial advances and retreats we've come to call the Ice Age. In fact, some flows apparently cooled against glacial ice and incorporated glacially deposited gravel. The research made it clear that these huge explosions were separate from the earlier volcanism that created Mount Holmes and the rest of the Absaroka Range. Different rock, different time. In fact, the explosions obliterated a vast swath of these earlier mountains.

The three caldera-forming eruptions, as brief and explosive as they were, were each part of a grand cycle lasting hundreds of thousand of years. Volcanic activity began with the development of a large subterranean magma chamber, the swelling of the land above the magma chamber, severe faulting around the magma chamber as the bedrock rose and stretched, and several eruptions that covered the land with rhyolitic lava. There were also some eruptions of oozing basaltic lava. Though spectacular by most standards, these eruptions were merely warm-ups for the main event. As the ring fractures grew, magma began to escape to the surface. Relieved of the pressure underground, gas frothed out of the viscous pitchlike magma (some 76 percent silica) violently and suddenly. This explosion fed on itself: As more magma frothed and escaped, the pressure in the magma chamber dropped, allowing the degassing of more magma. The reaction multiplied and built to a terrific explosion. The depleted magma chamber collapsed, the land dropping like a giant piston, forming the immense caldera. Then for hundreds of thousands of years following, intermittent eruptions filled much of the caldera with new lava flows. These subsequent eruptions, as recent as 70,000 years ago, form much of the topography we see today. Finally, the cycle led to resurgent uplifting of the land above the magma chamber—as appears to be happening today, and in the past, at least, that led to the next big event.

— . —

When Christiansen and Blank began their work, the theory of plate tectonics was in full eruption. New ideas about how the earth worked, though still controversial among a few steadfast holdouts, were boiling out of the geology. "We were very much in the thick of it," Christiansen recalls now, more than forty years later. At first the theory seemed to bear little relevance to their work. But "it didn't take very long before we realized there was a regional context." By 1968, he was spending a lot of his time not only in Yellowstone, but also out to the southwest, on the Snake River Plain.

Beginning in the 1930s, geologists began to recognize the similarity between the welded volcanic tuffs of Yellowstone and the rocks of the eastern Snake River Plain, a vast belt of fertile volcanic soils, abundant groundwater, and rhyolitic and basaltic formations increasingly given over to potato growing. In the late 1950s, Warren Hamilton pointed out the close relationship of quakes and volcanoes in Yellowstone and the Snake River Plain, especially the emplacement of similar rhyolitic and basaltic lava flows. But the source of these volcanic formations—that is, huge caldera-forming explosions—and why these Snake River formations should resemble Yellowstone still eluded geologists.

But the new theories suggested a link between Yellowstone and the region to the southwest. In 1963, Canadian geologist J. Tuzo Wilson had come up with a novel explanation of oceanic island and seamount chains such as the Hawaiian Islands. Ancient Hawaiians long recognized by patterns of erosion, soil, and vegetation that the northwestern islands, such as Kauai, appeared oldest. The islands became progressively younger to the southeast. Hawaiians explained the progression by many versions of the legend of Pele, goddess of volcanoes, who lived on Kauai. In one story, Pele was driven away by her sister Namakaokaha'i, goddess of the sea. Pele fled to the southeast, and each place she landed, volcanism commenced—first on Oahu, then Maui, and finally Hawaii. Wilson imagined the evolution of the island chain differently, but no less fancifully: He thought of someone lying on his back on the bottom of a quiet stream, blowing bubbles through a straw. The chain of bubbles bends in the slow current and emerges at the surface, in Wilson's theory, forming islands.

What Wilson proposed were thermal plumes, several permanent hot spots scattered throughout the oceans of the world. These hot spots originated deep in the hot mantle of the earth. Hot plumes burned upward, bending through the current of the upper mantle and bursting to the earth's surface as erupting volcanoes below the sea floor. These eruptions—the steady, oozing basaltic eruptions typical of the Hawaiian Islands—built heaps of lava as seamounts. The seamounts would build, rise, and eventually emerge above sea level as islands. As this was occurring, the oceanic plate moved over the hot spot at a nearly steady rate. Eventually the crust would move the newly created island beyond the reach of the hot spot, cutting off the supply of magma. Something, such as a thicker chunk of crust or an interruption in the supply of upwelling magma, would interrupt the flow as the crust moved along. But then the plume would find another weak spot and begin forming another seamount. This new seamount eventually grew, rose, and emerged above sea level as an island. And so on. The plume first formed Kauai, where the oldest volcanic rocks date to about 5.5 million years ago. Then, as the Pacific Plate moved steadily to the northwest, the hot spot created Oahu and then Maui. Now, the hot spot breaks to the surface at the big island, Hawaii, where the oldest rocks are about 700,000 years old and volcanoes still regularly ooze molten basalt. Similar hot spots form oceanic island chains throughout the world, such as the Azores and Galápagos.

With these ideas in mind, Christiansen and Blank published their first report on Yellowstone. They also laid out the progression of ages of volcanic rocks up the Snake River Plain—from the oldest in southwestern Idaho and northwestern Nevada along a path that leads to Yellowstone.

From the late 1960s, a flood of articles began laying out the theory of plate tectonics. In 1970, Tanya Atwater at the University of California proposed that much of the volcanic and mountain-building activity resulted from the crunching of boundaries between the North American and Pacific plates. By 1971, Princeton geologist Jason Morgan advanced the theory of about twenty plumes rooted in the deep mantle beneath the sea floor. Soon after, Richard Armstrong published the potassium-argon dating of volcanic lava flows in the Snake River Plain. "That was the beginning of our understanding of this northeast progression," Christiansen said.

---

*When this rhyolitic magma finally reached the surface through numerous faults, it did not flow. Instead, it triggered some of the biggest explosions in the recent history of the earth.*

---

Scientists were beginning to see that Yellowstone volcanism was related to a larger pattern—a much larger pattern. Yellowstone was hooked up to the very forces that shuttled the earth's tectonic plates over the surface of the globe. And the three caldera-forming explosions in Yellowstone were only the most recent in a long line of volcanic explosions. In 1972, Morgan suggested that Yellowstone and the Snake River Plain were tied together in a single system—a series of volcanic calderas fired off by a hot spot originating deep below the crust that triggered one caldera explosion after another as the North American Plate ground slowly to the southwest. The system worked pretty much like the Hawaiian Island chain. The difference was this: In Hawaii, the hot spot burned through basaltic crust, creating relatively mild-mannered eruptions of flowing lava. Beneath Yellowstone, the hot spot melted the underside of the continental crust, which was made of rhyolite. Molten rhyolite created a chamber of viscous magma three to four miles below the surface. When this rhyolitic magma finally reached the surface through numerous faults, it did not flow. Instead, it triggered some of the biggest explosions in the recent history of the earth.

— . —

Wilson's and Morgan's original hot spot theory envisioned a plume of molten material rising from deep within the earth, near the mantle-core boundary, about 1,700 miles deep. That model has come to be known as the mantle plume origin of hot spots. More than a hundred of these hot spots, including the one below Yellowstone, are responsible for driving the movements of plates and stretching the earth's crust. This stretching causes, among other things, the basin-and-range mountain building across much of Nevada and portions of Utah, California, Oregon, Idaho, and Montana. And many scientists still argue for it.

But Christiansen and many others reason that the origin of the Yellowstone hot spot lies in the upper mantle, perhaps only 125 miles deep. According to Christiansen's version of this theory, the interaction of the plates off the western coast of North America crunched the continental plate, pushing up the Cascade Range of volcanoes and mountains. But later movement relieved some of the compression, allowing the crust in the American West, including Yellowstone, to rebound and thin. At the same time, interactions between the continental plate and the oceanic plate sliding beneath it caused widespread melting of the upper mantle and an abundant source of hot magma beneath the crust. In Christiansen's view, the thinning of the crust created the hot spot, not the other way around: "This propagating wave of basaltic magmatism initiated the Yellowstone hotspot by crustal melting." This source of molten rock found weaknesses in the thinning crust— weaknesses that became the magma chambers and then the calderas of the Snake River Plain and finally Yellowstone. Champions of these competing theories wrangle over highly technical evidence in what Christiansen has called an ongoing "legitimate scientific dispute."

Robert B. Smith, professor of geology and geophysics at the University of Utah, has presented a somewhat different picture—that of a "beheaded" plume (which he calls a "plum" because it doesn't have a long stem leading deep into the mantle). Smith and his colleagues have relied on the many seismic stations in and around Yellowstone to track seismic waves from the area's frequent earthquakes. The waves travel at different speeds through rock of different temperatures and state of melt. From this information they have constructed three-dimensional images of the structures beneath the park. They believe they see a 50-mile-wide

The volcanic Cascade Range, extending 700 miles along the western coast of North America, formed from the subduction of the oceanic plate under the continental plate. Mount Hood, one of the stratovolcanoes of this range, rises 11,235 feet in northwestern Oregon. *James Saunders, Shutterstock*

plume of molten rock that originates at least 410 miles underground in the upper mantle and rises to within 75 miles of the caldera. There it encounters cooler rock and spreads out some three hundred miles wide. Why does the plume end about four hundred miles down? One explanation, says Smith, is that a plume originated deep within the mantle. But then as it rose into the conveyor-belt flow of the mantle, it was cut off from its roots.

"It doesn't matter to the earth's surface at Yellowstone where the magma and heat come from," Smith said in a recent talk at Yellowstone, where he and his audience sat just three or four miles above a still-active chamber of red-hot magma. "It doesn't care out here. It's a huge volcanic system. It's an active earthquake system. It doesn't matter if it's a deep plume source or a shallow decompression source. All the effects remain the same."

— · —

The exact origin of the hot spot aside, what do scientists know for sure about Yellowstone?

The three Yellowstone calderas are the latest in a long line of caldera-forming volcanoes that began erupting nearly 17 million years ago in what is now northwestern Nevada.

The hot spot track did not create a continuous series of volcanoes. Rather, it left a pattern similar to the Hawaiian Islands: Evidence of prolonged activity and then it would move on. The next center of volcanic activity would appear to the northeast. The Yellowstone hot spot created a series

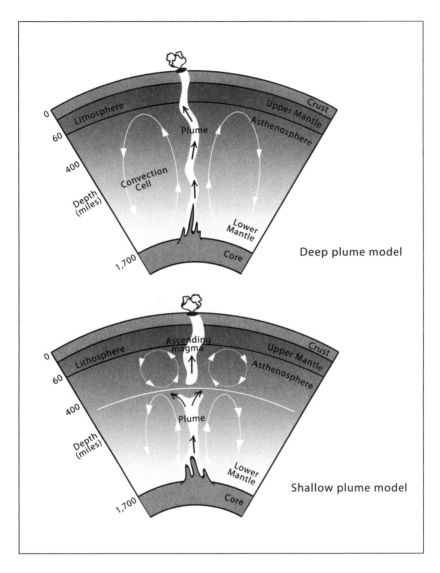

A rising plume melts the rhyolitic crust near the surface of Yellowstone, creating a chamber of viscous magma. As the molten rock escapes to the surface, relieving pressure in the chamber, gases escape with explosive power. Scientists continue to debate whether the hot spot originates deep within Earth's mantle (left) or from a shallower location (right). *Adapted from Robert B. Smith and Lee J. Siegel,* Windows into the Earth: The Geologic Story of Yellowstone and Grand Teton National Parks of caldera-forming explosions at each of several sites along this track: the McDermitt volcanic field along the Oregon-Nevada border 15–16 million years ago. The Owyhee-Humboldt field at the corner of Oregon, Idaho, and Nevada 13–15 million years ago. The Bruneau-Jarbidge field 10.5–12 million years ago. Twin Falls field 8.6–10.5 million years ago. Picabo field north of Pocatello 7–10 million years ago. Heise field 4.3–6.5 million years ago. And finally the Yellowstone area beginning 2.1 million years ago. Over 16.5 million years, the hot spot track traveled nearly 350 miles, at a speed of one to two inches a year.

In all, there have been at least 142 Yellowstone-scale eruptions along the track during the last 16.5 million years, according to recent research by University of Utah geologists Michael Perkins and Barbara Nash. Perkins and Nash tracked down ash fall layers, determined their chemistry, and developed a chronology of the eruptions that created them. (The scientists didn't include at least four large eruptions that occurred slightly earlier and left layers of different chemical composition, believing they may have been part of a different event.) Nash and Perkins also showed that the rate of eruptions has slowed over time. The hot spot started with a bang, literally, with more than thirty major eruptions per million years in the early going, gradually slowing to a rate of fewer than three major eruptions per million years during the past 8.5 million years.

The volcanic activity all along the eastern Snake River Plain resembled the process that would later occur at Yellowstone. As the crust moved over the hot spot, it bulged 300 miles wide and 1,700 feet high—like pulling a rug over an

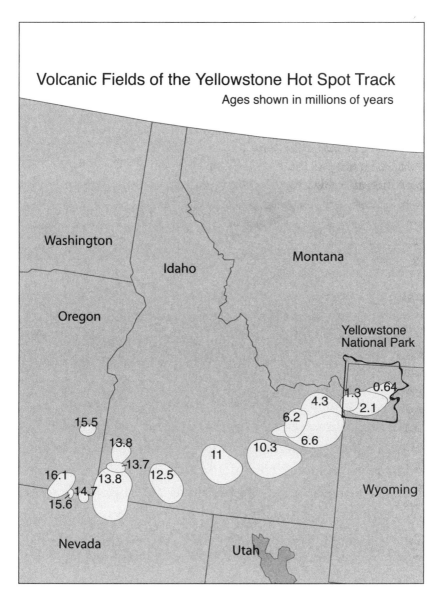

# Volcanic Fields of the Yellowstone Hot Spot Track

Ages shown in millions of years

For nearly 17 million years, as the North American Plate has moved slowly southwestward over the Yellowstone hot spot, the rising plume of magma has triggered a series of at least 142 caldera-forming eruptions across southern Oregon, northern Nevada, southern Idaho, and western Wyoming. The major volcanic fields, some containing many calderas, are depicted here.

object on the floor, as Robert B. Smith and Lee Siegel describe it in their book *Windows into the Earth*. With the stretching came a buildup of rhyolitic and basaltic eruptions, immense caldera-forming explosions, and a run-out of more rhyolitic and basaltic eruptions. As the crustal plate moved away from the hot spot, the caldera field returned to a lower altitude. (Think of that object under the rug.) The Yellowstone Plateau has been raised to an average 7,900 feet. In time, as the hot spot moves on, the plateau will probably fall 2,000 feet to the level of the rest of the Snake River Plain. The depression in the wake of the hot spot traces the concave edge of the tectonic parabola of faults and earthquakes. As the rhyolite of dying volcanic fields hardened, fractures spread and reached deep into the earth. Rhyolitic and basaltic magma erupted through the fractures. Extensive basaltic flows filled the calderas and created a level plain.

The volume of these flood basalts may be even greater than the volume ejected in catastrophic explosions. They left a landscape resembling Craters of the Moon National Park—vast fields of rough, ropey black rock, the cooled basalt, with lava tubes, craters, and cones of past volcanoes. The ground is covered by volcanic cinders, lightweight rock filled with the tiny pockets, called vesicles, left by gases in the frothing lava. And for the longest time, like Craters of the Moon today, the basalt lay barren. It took centuries for hardy pioneering plants to colonize the hostile ground—lichens, monkeyflower, dwarf buckwheat, lava phlox, bitterroot, scorpionweed, and limber pine.

The modern understanding of Yellowstone's formation

evolved quickly during the last forty years. Christiansen, the once-young Yellowstone geologist, retired in 2003. But he still has an office (now in Menlo Park, California). And he still comes into work to speculate, theorize, and write about the young volcanic rocks of the western United States, including Yellowstone. "Now," he said, "I can come in late and leave early."

# CHAPTER 6

# DISTANT DEATH

In the summer of 1971, paleontologist Mike Voorhies was digging in an eroding ravine that cut through a layer of prehistoric volcanic ash. He had been working his way deeper into the gulley when he spotted a skull—the skull, clearly, of a baby rhino that had lived, once upon a time, in northeastern Nebraska.

Voorhies had grown up in Orchard, just six miles away. On family picnics, he and his sisters had played in the creeks that cut down through the hills in these parts. There in the water they'd find petrified wood and fossilized teeth. Fossils were, as he described it, part of the "background noise."

He had gone away, but not far—to the University of Nebraska at Lincoln, to become a pharmacist. It seemed to him like a comfortable career. But then he took a geology course from Sam Treves, an Antarctic geologist and mineralogist, who inspired young Voorhies to believe he could study nature and actually make a living of it. At the end of his freshman year he saw a bulletin board notice offering—he describes it a bit facetiously now—"a dollar a day and all the road kill you could eat" to collect fossils for the University of Nebraska State Museum, one of the biggest in the Midwest.

So, years later, with his Ph.D., he was teaching at the University of Georgia, the only vertebrate paleontologist at the time in the state, a hotbed of creationism. And this summer he was back home in Nebraska with his wife, Jane, a geologist, to conduct a reconnaissance of possible paleontological sites in the Verdigre Creek valley.

You don't do geology in a landscape like this, a professor once told him. And he was right. It was covered with grasses, crops, and soil. You had to hunt to find bedrock, much less the fossils in the bedrock. It would be nearly a miracle to find something here.

But he had his degree. He had the summer to himself. He could do what he wanted. So he and his wife spent the long days driving around asking permission to hike around and peck at eroded banks and bedrock outcrops with a rock hammer.

They had come to Melvin Colson's farm. Colson grew corn on the flat high ground. But the rest was rangeland, gullied, with thickets of sumac and thin soils covered with prickly pear. The reason Voorhies had asked to explore the land was simple. It came down to a single formation: a high cliff of exposed bedrock, a sliver of naturally cemented river sand well known to geologists. Farther west, where the layer is better exposed, the layer had turned up fossil elephants. So at first Voorhies and his wife tapped away at the cap rock in the cliff. They found a few fossilized bones. And at the base of the cliff, they found a two-foot-thick layer of volcanic ash, right where it was supposed to be. The layer had been known for years, though no one knew exactly where the ash had come from—presumably from some grand eruption in the western United States millions of years ago.

Jane had quit the cliff after a few days—the base was covered in poison ivy, and she was sensitive to it. Voorhies finished his work there and began exploring exposures of the ash layer. It was moist and powdery fine. The gulley cut upward through the ash layer. He examined one side of the gulley, mapping in the details of the geology, and began working the other side, working deeper into the ravine when he saw about six inches of skull sticking out from the base of the ash. "I knew it was a baby rhino skull," recalled Voorhies recently. The species, *Teleoceras major*, the barrel-bodied rhino, was common on the Nebraska plains until it died out 5 million years ago. The following morning he poked around the skull and realized it was still attached to the neck. "I thought, my gosh, that's interesting." Unearthing the rhino was more than he was prepared to undertake at the moment, so he left it.

---

*The skull, the jaws, virtually the entire skeleton was lying in perfect articulation as if the animal had peacefully lay down to die and then was lovingly buried. "And outlining that first skeleton we found five more. Uh-oh. That's when bells started to go off."*

---

And there it sat. A memory and an × on a map. Georgia was a poor base to carry on field excavations a half continent away. But Voorhies finally came home, in 1975, to a job at the University of Nebraska. Two years later, he came out to the Colson farm with students and shovels. "We dug down on top and found out that the baby rhino skull was connected to a skeleton. Just like the old song says— The head bone connected to the neck bone. . . ." The skull, the jaws, virtually the entire skeleton was lying in perfect articulation as if the animal had peacefully lay down to die and then was lovingly buried. "And outlining that first skeleton we found five more. *Uh-oh*. That's when bells started to go off."

— . —

"There's no Eureka moment," Voorhies told me recently. "People sometimes say, 'Oh, you saw the baby rhino and your life changed forever.' Well, bullshit. That's not what happened. I suppose if I was smart enough I might have seen the potential of this little fossil, but I didn't. It was a very interesting fossil, but it was just another fossil."

But as time went by, and additional areas were excavated, the fossil and others buried with it became part of an increasingly intriguing find—a mass death site unlike any others. As work there continued, the fossils would shed light on a deadly aspect of ancient eruptions of the Yellowstone hot spot.

We were gazing across the prairie at the bedrock layer where Voorhies first began his explorations. The Colson farm is now Ashfall Fossil Beds State Historical Park. The farm and rangeland are slowly reverting to native prairie— waving grasses and coneflowers.

Voorhies is stocky, wears a blue shirt and slate gray dungarees, and has a farmer tan. His balding dome has left a fossilized band of sweat on his battered canvas cowboy hat. The brim is curled and crushed as if he has gripped it in his hands for as long as he has owned it, five seasons now.

"When did you realize what you had?"

"I suppose," he answered, "it was the point where we found that we had more than one rhinoceros."

Voorhies' team took photos of the rhino skeletons, including one that showed the skeleton of an unborn calf inside the skeleton of a pregnant cow. Voorhies sent them off to the National Geographic Society with a request for a research grant. Soon he got authorization to spend up to $25,000. He assembled his team and hired a bulldozer for the next summer. "Mr. Colson very graciously allowed us to mess up this area right here," Voorhies explained, tracing out the area—now excavated—where the hillside had covered the baby rhino and its mates.

"Before we came in and spoiled it with a bulldozer, this was a grassy hillside." Now we were sitting at the base of a twelve-foot cut. "See all the little sparkles?" Voorhies said. "That's pure volcanic ash. It's almost as soft as cigarette ash. It's about as fresh as it was 12 million years ago when it was blown out of the volcano."

That summer, the crew set to work, first with the bulldozer to remove the river sediments overlying the ash layer. Then, by hand, they excavated the ash layer itself. "Once you decide to do the digging, you stop being a scientist and you start being a manual laborer." The ash surrounding the fossils was soft, poorly cemented. "Our tool of choice for excavating fossils in this material is a very soft paintbrush. The classic idea of the paleontologist out there with his metal tools whanging away doesn't work here. This is almost cosmeticians' work."

Throughout most of the area, the ash layer lay as an even blanket, about two feet thick. But it soon became apparent that, in one area, the ash had filled in what had been a pond, in places more than ten feet deep.

The ash layer filling the pond was dense with fossils. Voorhies collected two hundred skeletons: the common barrel-bodied rhinos, three species of camels, three species of three-toed horses, two species of one-toed horses, a saber-toothed deer just eighteen inches at the shoulder, and a smattering of carnivores, including a fox-sized dog and raccoon dog, a crowned crane, a pond turtle, and a giant land tortoise. Some bones showed damage from scavenging by the hyenalike bone-crushing dog. All these animals were common to the area 12 million years ago; all are extinct now. Fossils of many of these species were also present above and below the ash layer, indicating they had lived there for some time and lived there again after the ash fall.

The ratio of carnivores to herbivores is what you'd expect to see in a watering hole on the Serengeti, Voorhies said. The lack of fish bones suggested the pond was not large or deep enough year-round for fish to survive. But the presence of pond turtles and frogs suggested the pond never completely dried up, at least not for long. The tortoises were up to four feet long and, if they at all resembled their modern relatives, lived only in frost-free areas.

The landscape must have been covered in grass, Voorhies noted: Fossilized stems and seeds of grasses lay within the herbivores' skeletons. The diggers found a few seeds of hackberry, which bore amazing resemblance to the seeds of the hackberries standing next to the visitor center today. But while the landscape we looked over is rolling, hilly, occasionally even rugged and dissected by streams, the ash layer occurs at a single elevation, indicating the landscape of 12 million years ago was level. River sediments suggested a floodplain. "As we reconstruct it, the area was probably as flat as the Bonneville Flats in Utah today," said Voorhies. A vast grassy plain marked by an apparently singular and rather deep water hole. "We believe that the reason there were so many skeletons in a small area is that this was a water hole that was very well known to the animals. . . . When we excavate the sand layer beneath the ash, there are hundreds of years' worth of bones accumulated basically one bone at a time, and trampled and broken and scavenged." It was the very same pattern Voorhies had seen on excavations of water holes in Africa.

So what had happened? "We had a nice functioning ecosystem here," Voorhies said. "Then the ash came in and spoiled it." At the time of the excavations, Voorhies had surmised a massive volcanic eruption somewhere in the western United States, possibly New Mexico, unleashed a cloud of ash that buried these animals alive. Voorhies wrote as much in a 1981 article for *National Geographic*. It was a stupendous scene of mass death, unrivaled among paleontological sites—a prehistoric Pompeii in which animals lived one moment, and then were overtaken by a cloud of death from the west the next.

— . —

But his initial interpretation of the disaster left Voorhies with unanswered questions.

First, what explosion could have been so large as to bury Nebraska in two feet of ash?

Several prehistoric eruptions were candidates for the ash fall. There had been major eruptions in New Mexico. Volcanoes formed the San Juan Mountains in southwestern Colorado. Perhaps it had been a predecessor to the Long Valley caldera in California. Bill Bonnickson at the Idaho Bureau of Mines suggested that the ash might have come from one of the early caldera-forming eruptions caused by

A watering hole was a magnet for large animals that lived on the flat plains of Nebraska nearly 12 million years ago, including now-extinct species of rhinos, horses, camels, and elephants. *Mark Marcuson, University of Nebraska State Museum*

the Yellowstone hot spot, when it was still beneath southwestern Idaho. By the mid 1990s, Mike Perkins, a geologist at the University of Utah, used the "fingerprinting" evidence provided by a mass spectrometer to date the ash to 11.83 million years ago and trace it to a huge caldera-forming explosion in the Bruneau-Jarbidge volcanic field, a major center of the Yellowstone hot spot, nearly a thousand miles west of Ashfall.

Second, and even more perplexing, was this: Why did the animals in the water hole seem to be sorted by size, with the biggest on top? "You never find a rhino skeleton under a horse," Voorhies said. "You always find the horses and camels underneath the rhinos." Voorhies might never have come up with the answer except for the observation of a veterinarian who dropped in at his paleontology lab.

The first summer Voorhies and his students removed skeletons, they noticed a coating on the long bones, a deposit Voorhies described as "spongy granular material." It was hard to remove. Voorhies had never seen anything quite like it and assumed it was some sort of mineral deposit on the fossilized bone. But one of his students preparing the fossils said the strange substance appeared to have once been bone, not a mineral coating.

"I finally had to admit, yeah, it really is, it's bone all right," Voorhies recalled.

Sometime later, a veterinarian stopped by the lab during a tour for museum visitors. "I know what that stuff is," he said. "That's Marie's disease."

"I had never heard of Marie's disease," Voorhies said. "But it turns out that the veterinary literature has information about this pretty rare condition. Animals that have suffocated over a period of a week or two, a new growth of bone will occur on pretty much all parts of the skeleton except the joints. It's the inverse of arthritis. Unlike arthritis, which can take years and years to develop on the bones, and is a very common thing to find in modern and fossil skeletons, Marie's disease is a very quick onset. When oxygen levels drop below a certain threshold the skeleton system begins to malfunction, and you get new bone deposited on places where it doesn't belong. This causes a high fever. The animals get very sick. I've seen pictures of horses that have had Marie's disease, and they are just ugly. They're completely immobilized with pain. Their legs are all swollen up. Their heads are swollen up. They can hardly move.

"The reason that the veterinarian immediately recognized Marie's disease when he saw it on our fossil is that he had actually done the autopsy on a group of horses that had been caught in a barn when it was on fire. They inhaled lots of smoke from hay. Lots of siliceous material had been in the air. They got the horses out of the barn. And they tried to keep them alive. They stayed alive for a week or so, but they swelled up and they got this high fever and they all died. This vet did the autopsy, because they were very expensive animals. It turned out that they had this stuff and were gone. That's how we learned there was such a thing as Marie's disease. Hypertrophic pulmonary osteodystrophy is the fancy name for this condition."

The diagnosis of Marie's disease helped explain the sorting of animals and opened up a new interpretation of the death scene.

One day, after years, centuries, or even millennia of preliminary earthquakes and lava flows, the bulging landscape under what is now southwestern Idaho—a landscape that, except for the horses, camels, rhinos, and elephants, looked much like Yellowstone today—let loose with a terrific blast. Entire mountains were vaporized. As molten rock exploded from the magma chamber, pyroclastic flows raced across the landscape at jetlike speeds. For miles, the forest was flattened and everything killed. Fires raged. The entire region was black with smoke. Volcanic ash, smoke, and gas rose in a towering column.

The ash was carried on the prevailing westerlies. The finest rose to forty thousand feet, entered the jet stream, and traveled eastward at a hundred miles per hour. As the plume of ash bent with the wind and sped along, chunks of volcanic glass rained down to earth. The heaviest particles fell out first. Then progressively finer particles fell to earth. Within about ten hours after the explosion began, a dark cloud sped over the Ashfall water hole, bringing persistent darkness. A blizzard of ash began to fall, as if it were snow. The ash fall consisted of jagged pieces of rock and volcanic glass, the particles just a few microns in diameter. By the end of day one, the ash lay nearly two feet deep, covering all but the tallest grasses and shrubs.

With the slightest wind, the fine ash would swirl and drift into the only topography on the plain, the watering hole. Within a day or so, the turtles would have buried deep into the mud below the ash to wait futilely for better times. Some of the birds might already have died.

The mammals were able to wade through the deep powder. But eating had become difficult. As days went by, the grazers inhaled more and more of the fine, abrasive particles, which were too fine to filter and settled deep in their lungs. Within days, the lungs of the saber-toothed deer and smaller camels were so filled with dust, their breathing was labored and ineffective. The ash was sharp and abrasive, and their lungs may have been filling with blood as well. Their limbs were swollen. Pain and discomfort drove them to the water hole.

By now, swirling ash had filled the pond to several feet deep. The smaller animals congregated in the shallow water, but they still breathed the fine ash. They began to die from respiratory and pulmonary distress. The larger animals were feeling the effects of labored breathing. They, too, began to stream into the water hole. In the final days in the life of the watering hole, the rhinos packed into the water, the bulls, cows, and calves all crowded together, trampling the bodies and breaking the bones of smaller camels and horses beneath their hooves. The young rhinos perhaps had been spared some of the effects of the ash because they nursed rather than rooted through the ash for grass and forbs. But finally, the adults went down in the water, even as the youngsters tried to suckle their mothers. Within two to three weeks after the eruption, all the herbivores on the plains had died. Bone-crushing dogs visited the water hole and tried to scavenge the ash-covered bones. But the carcasses were soon covered by several feet of drifting ash, not to be seen again for nearly 12 million years.

Sometime after the ash fall—after the river had flooded and spread sand and clay among the ash, a period of months, years, perhaps decades—a small herd of elephants walked along the shore of the pond, leaving their tracks.

— · —

"Mount St. Helens killed a lot of people but mostly it just buried them and fried them and so forth," Voorhies said. "Pompeii killed lot of people but that was close enough to the volcano that you actually had toxic gases and hot mud and all kinds of Hollywood effects. This stuff was just dry powder. And the fact that it happened to be volcanic in origin probably didn't have a whole lot to do with the fact that it killed the animals. Any micron-sized dust is a health hazard. And if we were manufacturing any material of that grain size, we would be closed down by the Environmental Protection Agency. This is just dangerous stuff. It turns out that these super volcanoes will deliver unimaginable volumes of very, very, very fine material like that."

*Presumably there are similar watering holes, similar death sites, scattered over hundreds of miles, awaiting discovery.*

"Every time they wanted to get a mouthful of grass, they would inhale it," Voorhies said. "We know they continued to feed because of grass in mouths. They would have to snuffle up a lung full of ash." Because the smaller animals had smaller lungs and higher breathing rates, "the little guys would fill up and die soonest."

All the mammals showed signs of Marie's disease. None died instantly.

"When the ash fell, it sort of stands to reason this is where the critters would come. For one thing, if you have Marie's disease, you have a high fever. The mud probably felt really good to them. We believe probably their last few days or weeks were spent kind of milling around and not having a clue as to what was happening to them. Calf rhinos and little horses stayed with their mothers. We've actually got cases of mother and calf right next together," he said.

"Probably within two, three weeks of the time that the ash first fell, everything on the landscape was dead." The ash continued to drift into the pond. The bottom three feet of ash is packed with fossils. The upper six feet has none. "We've never found a trace of life in it."

For a while, the landscape remained a ghostly scene of light gray fading to a gray horizon. In time, the sky cleared (though spectacular sunsets lingered for years). Eventually, rain soaked and compacted the ash. Floodwaters from the nearby river laid down layers of new sediment. "The first sign of life we see in the ash bed after the big kill," Voorhies said, "is a bunch of elephants came wandering across the top here. . . . The landscape started to come back to normal." Plants either managed to push up through the ash or to colonize it. Located above the elephant tracks are fossil plant roots. "At that point we start finding bones and teeth of the very same species that were involved in the mass kill." Animals from far away probably migrated to the water hole after the landscape greened up. The colonization may have taken years, decades, even centuries. "Geologically, they are absolutely instantaneous."

Though ash covered what is now the Great Plains, scientists don't know if the death scene at Ashfall was repeated throughout the region. Presumably there are similar watering holes, similar death sites, scattered over hundreds of miles, awaiting discovery. "The beauty of an ash fall deposit is that it gives you a snapshot of ancient ecology. . . . Most super volcanoes deal lots of death and destruction, but they also provide a future historian of the earth with a record of exactly what it was like underneath the ash cloud."

Today, the Ashfall visitor center and paleontology lab overlook the gully where Voorhies began his excavations. A path leads from the visitor center down a hillside, intersecting the layers of earth spanning the time since the Bruneau-Jarbidge eruption. This walk backward through time is interpreted with small signs: 2.4 million years ago: First glaciers reached eastern Nebraska. Giant tortoises died out in America. 4 million years ago: Round-tailed beavers reached over two hundred pounds. 5 million years ago: Extinction of all American rhinos. 8 million years ago: Shovel-tuskers were Nebraska's most common elephants. 9 million years ago: Saber-toothed deer died out in America. 10 million years ago: Nebraska alligators and giant salamanders died out. 11.8 million years ago: Ashfall.

*This is kind of a strange case. No active geology happening anywhere within a thousand miles of here, and yet, we're actually looking at the result of a truly cataclysmic eruption.*

The trail leads to the "rhino barn," an open-sided shelter thirty-two feet by sixty-four feet, covering a fraction of what was once the watering hole. Voorhies and other workers have partially excavated about forty rhinos (and a few horses, camels, and deer) exactly as they died, lying on their sides—old bulls, babies with their mothers. In a few cases, cows carry fetuses in their pelvic cavities. It is a riveting and

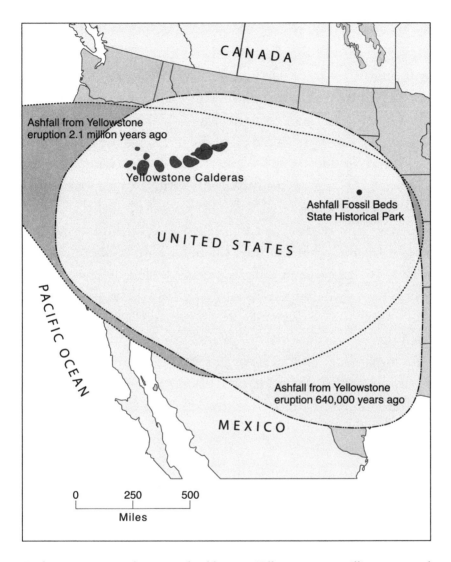

Explosive eruptions that created calderas in Yellowstone 2.1 million years and 640,000 years ago covered the western United States in volcanic ash several inches to several feet deep. A similar eruption nearly 12 million years ago blew a caldera in the Snake River Plain and covered the West in ash, killing animals now fossilized at Nebraska's Ashfall Fossil Beds State Historical Park. The patterns of ash fall depended on upper atmospheric winds at the time of the eruptions. *Adapted from Robert B. Smith and Lee J. Siegel,* Windows into the Earth: The Geologic Story of Yellowstone and Grand Teton National Parks literal representation of the animals' last moments, as though they had just lain down to sleep. Their bones look as though they have been carved from stone in high relief.

"Most fossils don't look like animals," Voorhies said. "I don't have to tell you that this was an animal. The volcanic ash was so fine it preserved all of the details of the creatures."

Voorhies likes the look of these rhinos lying now as they died. He regrets his initial excavations to remove skeletons from the hillside. "I wake up at night and I think, my god, you have vandalized this place. I have vandalized one of nature's masterpieces. As I think back on it, there was no reason to take those two hundred skeletons out of the ground. They belonged down there. But at the time, it was wide open to the elements and there was no plan for any kind of a park or any sort of protection for these fossils. But you look back and everybody's life probably has points where you think, man alive, if I had to do it over I'd do it a different way. I often wish somebody a lot smarter than me had worked on this place.

"But I enjoy what we're seeing now—a family coming down the hill and I think they're going to learn something. They're going to see what few people have really had a chance to see—that is, fossils as they really do occur in nature. Whether it will resonate with them, when they start seriously thinking about the history of the world, whether it all happened instantaneously or whether Darwin was right and there was some gradual change involved, I don't know. But I don't think it's going to hurt them.

"This is kind of a strange case," Voorhies said, "no active geology happening anywhere within a thousand

As ash fall from a super eruption on the Snake River Plain buried the Great Plains, animals succumbed to pulmonary disease and died in a shallow pond in northeastern Nebraska. Their skeletons, in their death pose, were discovered at what would later become Ashfall Fossil Beds State Historical Park. *University of Nebraska State Museum* miles of here, and yet, we're actually looking at the result of a truly cataclysmic eruption that occurred a thousand miles west of here and simply blanketed the landscape and ended up killing lots and lots and lots of animals. Not the way volcanoes usually kill things, which is blowing them up or poisoning them, the dramatic type of death, but simply choking them to death on extremely fine dust.

"If we can predict anything in geology, I think we can say the hot spot is not dead. Anyone can look at Yellowstone and know something is happening there."

As I drove away, back through the pin-neat prairie town of Orchard, I imagined the new-mowed lawns and parks, the Lunchbox Café, the Udder Place Bar, and Holz Lumber buried under two feet of white ash, no crops in the ground, domestic stock all dead, and the summers turned cold.

# CHAPTER 7

# MOST-SUPER VOLCANOES

Super volcanoes the size of Yellowstone are rare. But
there have been a few. Unlike Yellowstone, most have
gone dormant, leaving behind massive ash tuffs as testa-
ment to their power, but otherwise not calling attention to
themselves with earthquakes, geysers, or other geothermal
pyrotechnics. As a consequence, many have gone unno-
ticed, to be discovered only recently. Surely, some will come
to light in the future.

But first, a matter of definition: What is a *super* volcano? The term was coined only a few years ago for a BBC
science program on eruptions the scale of Yellowstone. At
the time, the term had no scientific currency. But recently,
geologists have been using the word to describe a volcano
of exceptional magnitude—the very biggest of which they
had been calling ultraplinian.

Defining *super volcano* begins with knowing how to
categorize the size of an eruption. Scientists have tried to
characterize volcanoes both by their magnitude (the volume or mass of magma ejected) and their intensity (the rate
this stuff is expelled). Did it spew out in days? Or just

moments? Those questions of time are hard to pin down. Scientists can read the tuff for signs that the rock cooled or even eroded between the emplacement of lava flows. But they can't tell if a blast occurred instantaneously or over a matter of hours or days.

Back in 1982, Stephen Self, now at the Open University in England, and Christopher Newhall at the U.S. Geological Survey devised a scale called the Volcanic Explosivity Index to rank the power of volcanic explosions, from tame Hawaiian-style basaltic flows to super-sized caldera-forming blasts like Yellowstone.

### VOLCANIC EXPLOSIVITY INDEX

| MAGNITUDE | DESCRIPTION | PLUME | EJECTA VOLUME | FREQUENCY | EXAMPLE |
|---|---|---|---|---|---|
| 0 | Non-explosive | <100 m | >1,000 m³ | Daily | Mauna Loa |
| 1 | Gentle | 100–1,000 m | >10,000 m³ | Daily | Nyiragongo (1994) |
| 2 | Explosive | 1–5 km | >1,000,000 m³ | Weekly | Unzen (1792) |
| 3 | Severe | 3–15 km | >10,000,000 m³ | Yearly | Nevado del Ruiz (1985) |
| 4 | Cataclysmic | 10–25 km | >0.1 km³ | 10s of years | Galunggung (1982) |
| 5 | Paroxysmal | >25 km | > 1 km³ | 100s of years | St. Helens (1980) |
| 6 | Colossal | >25 km | >10 km³ | 100s of years | Krakatau (1883) |
| 7 | Super colossal | >25 km | >100 km³ | 1,000s of years | Tambora (1815) |
| 8 | Mega colossal | >25 km | >1,000 km³ | 10,000s of years | Toba (72000 B.C.) |

At the quiet end of the scale—magnitude 0—are the nonexplosive eruptions like those on the island of Hawaii. These volcanoes can be immense. Mauna Loa, in fact, is the largest volcano on the planet. Its gradual flanks rise 2.5 miles above sea level and extend 3 miles below sea level to the ocean floor. Mauna Loa's great mass sinks into the sea floor an additional 5 miles. So in all, the summit rises more than 10 miles above its base—far higher than Mount Everest. Yet because its basaltic lava flows easily, it never builds up sufficient energy for an explosion.

Magnitude 1 are so-called gentle volcanoes, eruptions that expel ten thousand to a million cubic meters of ash and magma in an eruption lasting less than an hour and producing a column of ash generally less than a kilometer high. Impressive, certainly, if you happen to be in the vicinity, but unlikely to make the history books.

A magnitude 2 volcano on the VEI blows 1 million to 10 million cubic meters up to five kilometers high. Note that a magnitude 2 volcano is not twice as big as a magnitude 1. No, it is ten times as energetic. The VEI scale, like the Richter scale for earthquakes, is a logarithmic scale—each step a tenfold increase over the number before.

Some famous volcanoes rank rather middling on the VEI. The Mount Vesuvius eruption of A.D. 79 buried eight towns, including Herculaneum and Pompeii, in pyroclastic flows. It gassed, burned, and buried thousands. At the climax of the eruption, the column of ash stood twenty miles high. The Vesuvius eruption expelled a cubic mile of ash in less than a day. It rates magnitude 5.

Mount St. Helens, which erupted May 18, 1980, blew out the side of a mountain and killed fifty-seven people. It was one of the most momentous twentieth-century North American volcanic eruptions. The U.S. Geological Survey refers to it as the "worst volcanic disaster in the recorded history the United States." Yet it too rates only magnitude 5.

A whole order of magnitude more powerful was Krakatau in Indonesia, perhaps the most infamous volcano of all time. When it erupted in a paroxysm of fire, thunder, and smoke in 1883, it generated tsunamis and killed more than thirty thousand people. It rated magnitude 6.

Krakatau, as powerful as it was, doesn't count as a super volcano. Scientists such as Self have suggested the term be reserved for explosive eruptions that expel $10^{15}$ kilograms (roughly 500 cubic kilometers or a bit over 100 cubic miles) of magma. Such an eruption would be a high-level magnitude 7.

According to the U.S. Geological Survey, the term *super volcano* "implies an eruption of magnitude 8 on the Volcano Explosivity Index, meaning that more than 1,000 cubic kilometers (240 cubic miles) of magma . . . are erupted." In other words, two full steps above Krakatau on the VEI—one hundred times more powerful. And a thousand times more powerful than Mount St. Helens.

By either standard, two of the three most recent Yellowstone explosions would qualify.

Recently, volcanologist Ben Mason and colleagues from Cambridge University compiled a list of all known volcanoes with a magnitude of 8 or greater—forty-seven eruptions in all. (New super eruptions keep coming to light, and scientists often discover through additional field research that known eruptions turn out to be larger than once thought.) Mason jiggled the VEI a bit, estimating the density of expelled material to come up with "dense rock equivalent" instead of volume to distinguish densely welded tuff from lightweight cinder. And he fine-tuned the VEI to make his measurement of magnitude precise to a single decimal place. Most of the super volcanoes are located in western North

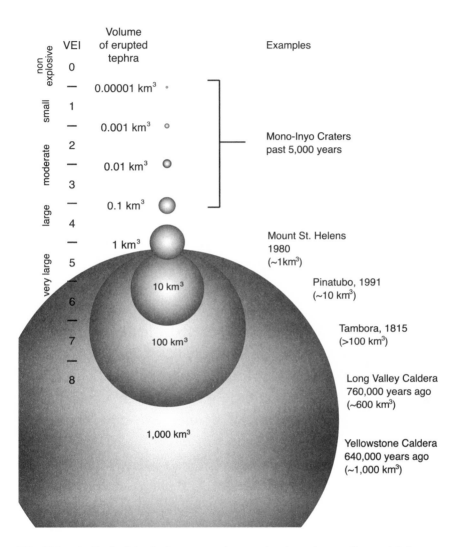

The Volcanic Explosivity Index rates volcanoes by the volume of material they eject. Like the Richter scale for earthquakes, the index is a logarithmic scale—each step a tenfold increase over the number before. The popular term *super volcano*, still in search of a scientific definition, has generally been used to describe volcanoes of magnitude 8 or larger. *Adapted from C.G. Newhall and S. Self, "The Volcanic Explosivity Index (VEI): An Estimate of Explosive Magnitude for Historical Volcanism,"* Journal of Geophysical Research *(1982)*

America, western South America, and Indonesia—generally places not far from subduction zones, where volcanics and mountain building have a long history of activity. But some also occurred in surprising places no longer associated with volcanoes or mountain building, such as Texas.

As a basis of comparison, the Huckleberry Ridge Tuff explosion at Yellowstone 2.1 million years ago rates a magnitude 8.7 on Mason's scale, ranking it fourth in size among known volcanic explosions. The Lava Creek Tuff eruption 640,000 years ago rated 8.3, placing it twenty-second. The third explosion at Yellowstone, creating the Mesa Falls Tuff 1.3 million years ago, even though larger than any eruption in modern time, was too small to make the list.

## LA GARITA, COLORADO

La Garita Caldera, located in the San Juan Mountains in southwestern Colorado, was created 27.8 million years ago in the greatest known volcanic explosion ever. The explosion, estimated at magnitude 9.1, ejected in a single cataclysmic eruption of 1,200 cubic miles of magma, enough to fill Lake Michigan to overflowing.

The full extent of the expelled material was discovered in the late 1990s. The Fish Canyon Tuff formed what geologists call a "single cooling unit," suggesting the explosion occurred quickly, layer upon layer piling up instantaneously without time in between to cool. The expelled magma was twice that ejected in Yellowstone's largest eruption, and some four thousand times the volume of the Mount St. Helens eruption. The collapse of the evacuated magma chamber created a caldera measuring more than twenty-one miles by forty-five

*The explosion, estimated at magnitude 9.1, ejected in a single cataclysmic eruption of 1,200 cubic miles of magma, enough to fill Lake Michigan to overflowing.*

miles. Scientists speculated an earlier nearby eruption, of the Pagosa Peak volcano, actually destabilized the La Garita magma chamber and triggered the much larger explosion.

Since the largest explosion of La Garita, seven additional major eruptions have occurred within the caldera. Among these was the magnitude 8.5 super volcano that created the Bachelor Caldera and the 290-cubic-mile Carpenter Ridge Tuff (ranked tenth among volcanic explosions) 400,000 years after the main explosion, just a blink of a geologist's eye. Then, about 27 million years ago, a magnitude 8.1 explosion (ranked thirty-third) created the San Luis Caldera and 135-cubic-mile Nelson Mountain Tuff. Then, 26 million years ago, the Creede Caldera expelled the 120-cubic-mile Snowshoe Mountain Tuff in a magnitude 8.1 eruption (number thirty-nine).

Since these eruptions, the resurgent caldera has lifted blocks of Fish Canyon Tuff more than four thousand feet thick to create the La Garita Mountains.

TOBA, INDONESIA

The eruption that created the caldera that now forms beautiful Lake Toba on the Indonesian island of Sumatra was the second-largest volcanic explosion known. It rated magnitude 8.8, blew out 670 cubic miles of ash and magma (12 percent bigger than Yellowstone's biggest). The caldera

# The Largest Explosive Eruptions on Earth

| Caldera Name | Deposit Name | Location | Caldera Diameter (km) | Volume (km³) | Magnitude | Age (million years) |
|---|---|---|---|---|---|---|
| La Garita | Fish Canyon Tuff | Colorado | 100×35 | 5,000 | 9.1 | 27.8 |
| Toba | Younger Toba Tuff | Indonesia | 100×30 | 2,800 | 8.8 | 0.074 |
| Unknown | Lund Tuff | Utah | N/A | 2,600 | 8.8 | 29 |
| Yellowstone | Huckleberry Ridge Tuff | Wyoming | 100×50 | 2,500 | 8.7 | 2.1 |
| La Pacana | Atana Ignimbrite | Chile | 60×35 | 1,600 | 8.6 | 4 |
| Unknown | Millbrig–Big Bentonite | Southeast U.S. | N/A | 1,509 | 8.6 | 454 |
| Unknown | Green Tuff + SAM | Ethiopia | N/A | 3,000 | 8.6 | 28–31 |
| Blacktail | Blacktail Creek Tuff | Snake River Plain | 100×60 | 1,500 | 8.5 | 6.6 |
| Emory | Kneeling Nun Tuff | New Mexico | 25×55 | 1,310 | 8.5 | 33 |
| Bachelor | Carpenter Ridge Tuff | Colorado | 20×28 | 1,200 | 8.5 | 27.5 |
| Timber Mountain | Timber Mountain Tuff–Rainier Mesa Member | Nevada | 25×30 | 1,200 | 8.5 | 11.6 |
| Paintbrush | Paintbrush Tuff–Topopah Spring Member | Nevada | 20 | 1,200 | 8.5 | 13.4 |
| Bursum | Apache Springs Tuff | New Mexico | 30×40 | 1,200 | 8.5 | 28–29 |
| Cerro Galan | Cerro Galan Ignimbrite | Argentina | 32 | 1,050 | 8.4 | 2.2 |
| Unknown | Kinnekulle Bentonite | Southeast U.S. | N/A | 972 | 8.4 | 454 |
| Bursum | Bloodgood Canyon Tuff | New Mexico | 30×40 | 1,050 | 8.4 | 28–29 |
| Unknown | Deicke Bentonite | Unknown | N/A | 943 | 8.4 | 454 |
| Uncompahgre | Dillon/Sapinero Mesa | Colorado | 20×23 | 1,000 | 8.4 | 28.5 |
| San Juan | Sapinero Mesa Tuff | Colorado | 22×24 | 1,000 | 8.4 | 28.5 |
| Paintbrush | Paintbrush Tuff–Tiva Canyon Member | Nevada | 20 | 1,000 | 8.4 | 12.9 |
| Chinati | Mitchel Mesa Rhyolite | Texas | 20×30 | 1,000 | 8.4 | 32–33 |
| Yellowstone | Lava Creek Tuff | Wyoming | 85×45 | 1,000 | 8.3 | 0.64 |
| Timber Mountain | Timber Mountain Tuff–Ammonia Tanks Member | Nevada | 25×30 | 900 | 8.3 | 11.4 |

| Caldera Name | Deposit Name | Location | Caldera Diameter (km) | Volume (km³) | Magnitude | Age (million years) |
|---|---|---|---|---|---|---|
| Porsea | Oldest Toba Tuff | Indonesia | 100×30 | 820 | 8.3 | 0.79 |
| Kilgore | Kilgore Tuff | Snake River Plain | 60×80 | 800 | 8.3 | 4.3 |
| Cerro Panizos | Panizos Ignimbrite | Argentina/Bolivia | 18 | 652 | 8.2 | 6.1 |
| Unknown | Barrel Springs Formation | Texas | N/A | 675 | 8.2 | 36 |
| Unknown | Wild Cherry Formation | Texas | N/A | 675 | 8.2 | 36 |
| Pastos Grandes | Sifon Ignimbrite | Bolivia | 40×50 | 1,200 | 8.2 | 8.3 |
| Unknown | Huaylillas Ignimbrite | Bolivia | 10 | 1,100 | 8.1 | 5 |
| Platoro | La Jara Canyon Member | Colorado | 18×22 | 592 | 8.1 | 30 |
| Taupo | Oruanui | New Zealand | 35×25 | 1,170 | 8.1 | 0.0265 |
| San Luis | Nelson Mountain Tuff | Colorado | 18 | 562 | 8.1 | 27 |
| Cerro Galan | Real Grande and Cueva Negra | Argentina | 32 | 510 | 8.1 | 4.2 |
| Turkey Creek | Rhyolite Canyon Formation | Arizona | 21 | 500 | 8.1 | 25 |
| Tucson Mountain | Cat Mountain Rhyolite | Arizona | 25 | 500 | 8.1 | 73 |
| Long Valley | Bishop Tuff | California | 20×35 | 500 | 8.1 | 0.7 |
| Minarets | Unamed | California | N/A | 500 | 8.1 | 100 |
| Creede | Snowshoe Mountain Tuff | Colorado | 24 | 500 | 8.1 | 27 |
| Mount Hope | Masonic Park Tuff | Colorado | 15 | 500 | 8.1 | 29 |
| Ute Creek | Ute Ridge Tuff | Colorado | 8 | 500 | 8.1 | 29 |
| Twin Peaks | Challis Creek Tuff | Idaho | 20 | 500 | 8.1 | 45 |
| Cowboy Rim | Gillespie Tuff | New Mexico | 18×26 | 500 | 8.1 | 33 |
| Juniper | Oak Creek Tuff | New Mexico | 25 | 500 | 8.1 | 35 |
| Organ | Cueva Soledad Rhyolite | New Mexico | 16 | 500 | 8.1 | 32 |
| Socorro | Hells Mesa Rhyolite | New Mexico | 25×35 | 500 | 8.1 | 33 |
| Blue Creek | Blue Creek Tuff | Snake River Plain | 30×55 | 500 | 8.1 | 5.6 |

Adapted from Ben Mason et al., "The Size and Frequency of the Largest Explosive Eruptions on Earth," Bulletin of Volcanology, (2004) measures more than sixty by eighteen miles and is nearly a mile deep. Toba's ash covered the eastern Indian Ocean and Bay of Bengal, an area at least half the size of the United States, reaching central Asia and the Middle East.

Toba's big blast occurred just 74,000 years ago, making it one of only two super volcanoes that may have been experienced by modern man's immediate predecessors— the biggest bang our kind has ever heard.

In fact, our species may just barely have survived it. The volcano spewed immense amounts of dust and sulfur dioxide. The gas combined with water to form fine droplets of sulfuric acid high in the atmosphere. These buoyant aerosols blocked and reflected sunlight. The volcanic winter that followed the eruption probably lasted for several years, depressing the average global temperature six degrees Fahrenheit or more. That is more dramatic than it sounds, with much greater cooling during summers at high latitudes. Cooling might have killed off both tropical plants and temperate forests. Greenland ice cores show a spike of carbon, suggesting that a widespread die-off of plants reduced carbon storage. In fact, according to oceanographers from the University of Rhode Island, Toba's volcanic winter might have triggered the resumption of continental glaciation during the most recent Ice Age (though others disagree that the volcano could have exerted so much influence on long-term climate).

Human geneticists studying mitochondrial DNA trace a lack of diversity in the human genome to a genetic "bottleneck"—that is, a period when the human population seems to have fallen to only a few thousand before rebounding.

They trace this bottleneck to a time that coincides with the Toba eruption. Stanley H. Ambrose of the University of Illinois at Urbana-Champaign proposed in 1998 that atmospheric and climate effects from the Toba eruption caused the coldest millennium of the Ice Age and brought widespread famine. The worldwide human population fell to 15,000 to 40,000 in isolated groups—"levels low enough for evolutionary changes, which occur much faster in small populations, to produce rapid population differentiation," Ambrose wrote in the *Journal of Human Evolution*. "Then modern human races may have diverged abruptly, only 70,000 years ago."

Toba's mega explosion 74,000 years ago was not the first volcanic activity in what is now the Toba caldera—or even the first super volcano. There were at least three earlier major eruptions. About 788,000 years ago, a previous Toba eruption, estimated at magnitude 8.3, ejected two hundred cubic miles of ash and magma. Mason and his crew place it twenty-fourth on the list of all-time biggest eruptions.

*Toba's big blast occurred just 74,000 years ago, making it one of only two super volcanoes that may have been experienced by modern man's immediate predecessors—the biggest bang our kind has ever heard.*

Nor does it appear that Toba has quite finished. Though it has not erupted in historical times, Toba did produce major earthquakes in 1892, 1916, 1920–22, and 1987. The resurgent caldera, formed by pressure from magma below, now forms Samosir Island and the Uluan Peninsula in Lake Toba.

Indonesia is one of the most seismically active areas in the world. Its islands form part of the Sunda Arc, where the Indo-Australian Plate, moving northeast, slides beneath the eastward-moving Eurasian Plate. In recent times, this subduction zone has produced the famous Krakatau eruption of 1883 and the deadly tsunami of 2004, which killed 300,000 people in Indonesia, Thailand, Malaysia, Bangladesh, India, and Sri Lanka.

## LUND TUFF, UTAH

A blast even larger than Yellowstone created the huge Lund Tuff in Utah, in the Great Basin province. The magnitude 8.8 volcano erupted about 29 million years ago. Like the even larger Fish Canyon formation to the east, the Lund Tuff forms what scientists have prosaically called a "monotonous intermediate ignimbrite." *Ignimbrite* is the stuff that flies from the volcano and then hardens (that is, the tuff and nonwelded rock). *Intermediate* because it is dacite, with less silica than rhyolite but more than andesite and far more than basalt. And *monotonous* because it's all the same. The Lund Tuff is a single cooling unit—it blew out of the volcano and hardened all in one short-lived event. Its volume may be as great as 720 cubic miles (which would make it even larger than Toba). The explosion appears to have created the White Rock caldera, about thirty miles across.

The Lund Tuff is part, albeit an awfully big part, of something called the Indian Peak volcanic field, astride the Utah-Nevada border. Numerous volcanic eruptions 27 million to 32 million years ago created dozens of calderas and blew out more than 12,000 cubic miles of tuff, twice the volume of all the Great Lakes combined.

VILAMA CALDERA, ARGENTINA

The immense Vilama Caldera in Argentina came to light too recently to make Mason's list. The caldera is associated with two mammoth explosions, the first about 10 million years ago and the second, larger eruption about 8.4 million years ago, according to fieldwork by geologist Miguel M. Soler of Argentina's National University of Jujuy. The second event formed two cooling units, suggesting it unfolded in two stages. In all it expelled five hundred cubic miles of material, ranking it near Yellowstone's largest, with a magnitude in the mid 8s. The depleted magma chamber collapsed, forming a caldera twenty-one miles by nine miles. Vilama Caldera's magma may have been the result of the grinding of the South American Plate as it overrode the oceanic Nazca Plate along the Pacific coast. The alternate compression and stretching of the crust produced large pockets of magma, Soler proposed.

The caldera is one of several in the Altiplano Puna Volcanic Complex on the central Andean Plateau, near the borders of Argentina, Bolivia, and Chile. Several, according to Soler, may qualify as super volcanoes.

LA PACANA, CHILE

La Pacana Caldera, another volcano in the Altiplano Puna Volcanic Complex, erupted some 4 million years ago in a magnitude 8.6 explosion that expelled 380 cubic miles of magma and created a caldera measuring thirty-seven by twenty-one miles. A thick tuff in the caldera floor uplifted to form the 16,000-foot Cerros de La Pacana.

MILLBRIG-KINNEKULLE BENTONITE
An oceanic plate, dragging island arcs or "microplates," collided with and sank beneath the southeastern coast of the United States 454 million years ago. The plate subduction triggered immense volcanic activity that left behind beds of K-bentonite—potassium-bearing clay formed from volcanic ash—in the rock record. These Ordovician eruptions laid down up to 60 ash beds in North America and 150 ash beds in what is now southern Scandinavia. Among the largest and most widespread are the Millbrig in eastern and central North America and the Kinnekulle in southern Scandinavia.

The Millbrig is a layer of bentonite one to two yards thick. The depth before compaction was probably four times as great. In Europe, the Kinnekulle forms a similar layer. In recent years, researchers have determined that these volcanic layers are the same age and may have come from a common source. If so, the volcano produced a dense rock equivalent of more than six hundred cubic miles—a magnitude 8.6 eruption. The Millbrig-Kinnekulle volcano apparently laid down these ash deposits in several closely spaced eruptions.

The volcano would rank not only as one of the largest super volcanoes but also the oldest. Unfortunately, we don't know exactly where it was. Warren Huff of the University of Cincinnati suggests that contours of the depth of ash converge on the present-day coast of North Carolina, South Carolina, or Georgia: "In Ordovician time this might well have been either an island arc or a magmatic arc along the plate margin. So although we can't say exactly where the volcano was, we do have a pretty good idea of its approximate location. Scandinavia and ancestral North America were much closer together at that time."

Another Ordovician volcano laid a similar thick bentonite ash bed, called the Deicke, across eastern and central North America—to all appearances in a single far-reaching explosion. The ash layer is equivalent to 220 cubic miles of dense rock, a magnitude 8.4 eruption, and the seventeenth-largest on Mason's list.

## GREEN TUFF

An unknown volcano somewhere in Yemen let loose with a magnitude 8.6 explosion about 30 million years ago, leaving behind the 720-cubic-mile Green Tuff in Yemen and Ethiopia. The eruption was part of paroxysm of activity between 29 and 31 million years ago related to the rift that formed the Red Sea and the Gulf of Aden.

## BLACKTAIL CREEK, IDAHO

As the North American Plate marched slowly southwestward, the Yellowstone hot spot let loose with thunderous eruptions in what is now Idaho, explosive footsteps up the eastern Snake River Plain. The hot spot would work away at a particular weakness in the crust, causing several caldera-forming explosions and numerous lesser lava flows in a volcanic field before moving on.

The Heise Volcanic Field in the eastern Snake River Plain is one of these regions. A bit older than Yellowstone, it lies to the southwest, just north of present-day Idaho

Falls. Three eruptions in this region rivaled the Yellowstone eruptions for violent power.

A magnitude 8.5 eruption 6.6 million years ago created the 360-cubic-mile Blacktail Creek Tuff. It ranks eighth among Mason's big explosions. A magnitude 8.1 explosion 400,000 years later formed the 120-cubic-mile Blue Creek Tuff (ranking forty-seventh). Then, 4.5 million years ago, a magnitude 8.3 explosion blew out the 190-cubic-mile Kilgore Tuff (ranking twenty-fifth).

Clearly, the hot spot now lurking beneath Yellowstone has released an unimaginable amount of energy during the last several million years.

EMORY, NEW MEXICO

The ninth-largest super volcano on Mason's list erupted 34.9 million years ago, creating the Emory Caldera and emplacing the 314-cubic-mile Kneeling Nun Tuff. The magnitude 8.5 explosion was the largest of dozens in southwestern New Mexico from 24 to 36 million years ago. Some of the most picturesque remnants of this cataclysm are visible at City of Rocks State Park.

---

*Humans undoubtedly experienced the volcanic winter and most likely even heard the explosion. Quite possibly, thousands living along the coasts were killed in giant tsunamis.*

---

LAKE TAUPO, NEW ZEALAND

The eruption that created the caldera now cradling New Zealand's Lake Taupo is the most recent of the world's super volcanoes. The largest blast, a magnitude 8.1 explosion just 26,500 years ago, blew out 127 cubic miles of magma, creating tuff more than three hundred feet deep in parts of the central North Island and a caldera measuring twenty-one by fifteen miles. It ranks thirty-second among the world's volcanic explosions. All evidence suggests that humans had not settled New Zealand at the time. But they did live in much of the rest of the world, including Southeast Asia and Australia. They undoubtedly experienced the volcanic winter and most likely even heard the explosion. Quite possibly, thousands living along the coasts were killed in giant tsunamis.

The Taupo volcano had erupted at least five times before. And even after the big blow, it wasn't yet done. In the last 26,000 years, it has awakened with twenty-eight major eruptions. It exploded again in about A.D. 181—in late summer according to evidence preserved in trees at Pureora Forest. The eruption blew out a vent near the Horomatangi Reefs, submerged now on the eastern shore of Lake Taupo, and continued for days or weeks. At its height, it blew out about seven cubic miles of ash and magma at speeds of several hundreds of miles per hour. There has not been a more explosive eruption in at least five thousand years. New Zealand was still unsettled; Maoris wouldn't reach the islands for nearly a thousand years. Today, the eruption would wipe out an area inhabited by more than 200,000.

At the moment, Taupo appears to be at rest. The Institute of Geological and Nuclear Sciences monitors the volcano with lake level records and a network of seismometers. Earthquake swarms stem from faults and continuing subsidence and crustal stretching rather than the movement of magma. Nonetheless, the Taupo Caldera is still active enough for New Zealanders to generate about 4 percent of their electricity from a geothermal plant near the lake.

## LONG VALLEY, CALIFORNIA

Geologists are particularly interested in the magnitude 8.1 explosion that formed Long Valley Caldera in east-central California 730,000 years ago because, like Yellowstone, the super volcano continues to show signs of activity.

The eruption on the eastern slope of the Sierra Nevada created the 120-cubic-mile Bishop Tuff. Ash from the explosion covered southern California, Arizona, New Mexico, Nevada, Utah, Wyoming, and western Kansas and Nebraska. The ejection of so much magma left behind a caldera measuring a mile deep and twelve by twenty-one miles. The main explosion ushered in hundreds of smaller eruptions and the creation of 11,050-foot Mammoth Mountain on the caldera wall. Replenishment of the magma chamber lifted the caldera floor. About 35,000 years ago, activity moved northward to Mono Lake to build an agglomeration of tuff, domes, cones, and craters known as the Mono Craters.

Activity has continued in historic times. Negit Island in Mono Lake is a black cinder cone that first erupted and emerged about 1,600 years ago. Panum Crater erupted barely six hundred years ago. Paoha Island is lake sediment pushed to the surface less than three hundred years ago. Fumaroles continue to steam on the island today.

An earthquake swarm, including four magnitude 6 quakes, ripped through Long Valley in May 1980. The activity signaled the onset of new volcanic activity, including the uplift of the caldera floor by about a foot. Earthquake swarms, thermal springs, and gas emissions continue to this day. Carbon dioxide gas, as much as 150 tons per day, seeps out of the ground, killing trees over a broad area. The U.S. Geological Survey has set up the Long Valley Observatory under its Volcano Hazards Program to monitor the activity and warn, if possible, of any new hazardous eruptions.

— · —

Some of the largest volcanoes never explode. They effuse basalt, which flows too easily to contain the energy necessary for an explosion. Instead, basaltic magma floods to the surface and produces huge mountains such as Hawaii's Mauna Loa and Mauna Kea. The output over time dwarfs the tuff from most known explosive volcanoes. Mauna Loa, for example, has produced an estimated 19,000 cubic miles of lava.

Two of the largest volcanic eruptions in the planet's history produced little in the way of volcanic explosions. But they were huge and profoundly affected Earth's climate and life. They coincided with two of the greatest extinction events in the planet's history.

The Siberian Traps, contrary to the sound of its name, was not a Paleolithic device to capture woolly mammoths but a flood basalt (resembling the more familiar Columbia River basalts of the northwestern United States). The eruption transformed and buried a major portion of Siberia. It was the greatest closely spaced series of on-land volcanic eruptions ever known.

Explosive super volcanoes have erupted at several rhyolitic volcanic fields around the world. Super eruptions have been most frequent in Yellowstone and the Snake River Plain, in the southwestern United States, and in the central Andes. *Adapted from Ben G. Mason, David M. Pyle, and Clive Oppenheimer, "The Size and Frequency of the Largest Explosive Eruptions on Earth,"* Bulletin of Volcanology *(2004)*

*Traps* derives from the Swedish word for "stairs," in reference to the steplike hills that form in these kinds of eruptions. Beginning about 248 million years ago, a hot spot burned through the thick, normally stable portion of the earth's crust known as the Siberian craton. The eruption began near the present-day city of Norilsk. For nearly a million years, the volcanism spewed from 380,000 to a million cubic miles of magma over an area perhaps as great as 3 million square miles. That's an area nearly the size of the contiguous United States in the heart of Siberia. The lava piled up as deep as four miles. If it had spread evenly around the earth, we'd all be buried by at least ten feet of Siberian rock. The traps, much eroded since they formed, now cover a plateau lying mostly east of the Yenisey River.

---

*The lava piled up as deep as four miles. If it had spread evenly around the earth, we'd all be buried by at least ten feet of Siberian rock.*

---

Scientists have suggested the Siberian Traps developed from a plume of magma originating deep within the earth's mantle. But this hypothesis has proved controversial, just as it has with Yellowstone's hot spot. Critics of the deep-mantle origin argue that plate collisions and subduction created a much shallower source of magma.

The eruption of the Siberian Traps corresponds with the Permian extinction, the most severe extinction event known. An estimated 95 percent of all marine life and 70 percent of land families vanished. The Permian period marked the confluence of continents into the supercontinent Pangea. This tectonic shifting also created a single open ocean, Panthalassa. The consolidation of landmasses and resulting changes in the world's single ocean might, in themselves, have contributed to the Permian extinction.

The movement and crashing of tectonic plates brought tremendous volcanism, including, quite possibly, formation of the Siberian Traps. These volcanoes released quantities of ash, carbon dioxide, sulfurous compounds, and other gases for hundreds of thousands of years. Even without explosions, the volcanoes' heat was enough to push materials high into the atmosphere. Ash and sulfuric-acid aerosols high in the atmosphere would have blocked and reflected sunlight, causing long-term cooling. At the end of the Permian, the sea level dropped lower than it has ever been, which might have occurred because huge continental glaciers tied up available water.

The eruptions might have caused other climate effects, some of which are at odds with global cooling. Prolonged and elevated levels of the greenhouse gases carbon dioxide and methane might have sent temperatures spiraling. Either prolonged warming or persistent cooling could have destroyed plant and animal communities on land and in the sea. There's some evidence that changes in the atmosphere caused large sections of the ocean to become starved for oxygen. Erupted chlorine and fluorine gases may have diminished the ozone layer, allowing ultraviolet rays to penetrate the atmosphere and kill organisms.

Recently, scientists have found evidence of major asteroid strikes at the close of the Permian. The energy and dust released from such collisions may have exacerbated the effects of the Siberian Traps. Luann Becker of the Department of Earth and Space Sciences, University of Washington, has proposed that "multiple catastrophes" might have been necessary to cause the most disastrous extinction in the earth's history.

— . —

While the causes of the Permian extinction are murky, it's accepted as a likely possibility that the impact of an asteroid accompanied the famous extinction of the dinosaurs at the close of the Cretaceous period 65 million years ago. But the deadly effects of the asteroid may have been aggravated by another massive episode of flood basalt volcanism, this time in west-central India, surrounding present-day Mumbai.

In 1980, physicist Luis Alvarez and others posited the most popular theory to date to explain the death of the world's terrestrial dinosaurs, flying reptiles, and large sea reptiles: An asteroid several miles across slammed into the earth, raising such a cloud of dust that the earth fell into a long winter. The evidence was compelling. A layer of iridium—a rarity in the earth's crust but common in asteroids—is found worldwide at layers corresponding to 65 million years ago. Cretaceous fossils are found below the layer, but never above. And an impact crater more than a hundred miles across, discovered on and offshore of Mexico's Yucatan Peninsula, dates to 65 million years ago.

But there is also compelling evidence that a simultaneous earth-shaking event contributed to the extinction. The

Deccan Traps volcanoes spilled as much as a half million cubic miles of basalt over an area equal to half of present-day India. In places, the basalt lies more than a mile deep.

Controversy continues over the duration of the Deccan eruptions. Scientists initially thought this massive volcanism occurred quickly—in a few tens of thousands of years. More recent dating suggests that eruptions continued for nearly a million years—perhaps much longer.

Early theories that the Deccan Traps resulted from a deep mantle plume have been attacked by scientists who favor the theory that grinding tectonic plates associated with the separation of the Seychelles microcontinental plate from India created a shallow hot spot that fired off the volcanism.

Regardless of the duration and cause, the Deccan Traps volcanoes were so large and persistent that they must be considered a factor in climate change at the time the dinosaurs disappeared. Whether they played the major role, or simply contributed to an environment already rendered uninhabitable by a speeding asteroid, is an issue scientists will continue to investigate and debate.

# CHAPTER 8

# THE DEADLIEST VOLCANOES

If indeed a super volcano was responsible for the Permian extinction, or wiping out the dinosaurs, or nearly exterminating the nascent human race—well, you can't find a catastrophe much deadlier than that.

But humans really haven't suffered much by super volcanoes, because, except for the Taupo explosion in New Zealand 26,500 years ago, physiologically modern man has not been around for any of the really big events. Even when Taupo blew, according to the evidence we have, humans weren't anywhere within a thousand miles.

Which isn't to say volcanoes aren't deadly. They are. In our short history on this planet, we have died by the hundreds of thousands in pyroclastic flows, mudslides, poison gas assaults, tsunamis, and secondary effects of volcanic eruptions. An estimated 300,000 people have died in volcanic eruptions just since A.D. 1600. Volcanoes, in fact, are tailor-made to kill human beings.

Most volcanoes are byproducts of subduction zones where oceanic plates scrunch beneath continental plates. So they tend to form on seacoasts, where most of the world's population lives. Worse, volcanoes usually don't form directly on the waterfront. Instead they arise on the inland side of broad, fertile coastal plains—the very places humans farm, settle, and build cities.

Volcanoes remain active for thousands or tens of thousands of years. So they are a long-term threat. They erupt often enough to be menacing but infrequently enough that people decide they can risk living nearby. So when they do go off, they take out unsuspecting victims by the village-full. The deadliest volcanoes in human history were surrounded by people who, by dint of human nature, chose to overlook the danger in their midst. But none of these volcanic killers—not one—was powerful enough to be called a super volcano.

Nowadays, we understand volcanoes much better than we once did. We have sensitive scientific equipment to detect the signals of an impending eruption. But when you get down to it, we are measuring the same old signs our Paleolithic ancestors responded to. If the mountain is smoking, the ground is shaking, hot lava is approaching, it's time to go.

## Tambora

More people have died from volcanic eruptions in Indonesia than in any other part of the world. The island nation has all the right ingredients for recurring volcanic disaster: dense population settled along a coast, atop the most seismically active zone in the world, the volcanic arc that runs parallel to the Sunda Trench (also called the Java Trench). There the Indo-Australian Plate, racing northeast at three

# THE DEADLIEST VOLCANOES

| DEATHS | VOLCANO | WHEN | MAJOR CAUSE OF DEATH |
|---|---|---|---|
| 117,000 | Tambora, Indonesia | 1815 | Starvation |
| 36,417 | Krakatau, Indonesia | 1883 | Tsunami |
| 29,025 | Mt. Pelée, Martinique | 1902 | Ash flows |
| 23,000 | Nevado del Ruiz, Colombia | 1985 | Mudflows |
| 14,300 | Unzen, Japan | 1792 | Volcano collapse, tsunami |
| 9,350 | Laki, Iceland | 1783 | Starvation |
| Unknown | Santorini, Greece | 1600 B.C. | Ash flows, tsunami |
| 6,000 | Santa María, Guatemala | 1902 | Ash flows, mudflows |
| 5,160 | Kelut, Indonesia | 1919 | Mudflows |
| 4,011 | Galunggung, Indonesia | 1822 | Mudflows |
| 3,500 | Vesuvius, Italy | 1631 | Mudflows, lava flows |
| 3,360 | Vesuvius, Italy | 79 | Ash flows and falls |
| 2,957 | Papandayan, Indonesia | 1772 | Ash flows |
| 2,942 | Lamington, Papua N.G. | 1951 | Ash flows |
| 2,000 | El Chichon, Mexico | 1982 | Ash flows |
| 1,680 | Soufriere, St. Vincent | 1902 | Ash flows |
| 1,475 | Oshima, Japan | 1741 | Tsunami |
| 1,377 | Asama, Japan | 1783 | Ash flows, mudflows |
| 1,335 | Taal, Philippines | 1911 | Ash flows |
| 1,200 | Mayon, Philippines | 1814 | Mudflows |
| 1,184 | Agung, Indonesia | 1963 | Ash flows |
| 1,000 | Cotopaxi, Ecuador | 1877 | Mudflows |
| 800 | Pinatubo, Philippines | 1991 | Roof collapses and disease |
| 700 | Komagatake, Japan | 1640 | Tsunami |
| 700 | Ruiz, Colombia | 1845 | Mudflows |
| 500 | Hibok-Hibok, Philippines | 1951 | Ash flows |

*Adapted from Russell Blong,* Volcanic Hazards: A Sourcebook on the Effects of Eruptions inches a year, is overtaking and sliding under the Eurasian Plate, moving eastward an inch a year.

The Ring of Fire around the Pacific is made of several subduction zones, where oceanic plates slide beneath continental plates. As the oceanic plate grinds downward, it causes deformation of both plates in the high-pressure and high-temperature environment of the upper mantle. Rock melts and this magma slowly works its way through crustal faults toward the surface, creating a volcano some distance above and away from the subduction zone.

The Sunda Trench stretches two thousand miles from the northwestern tip of Sumatra, along the southern edge of the Indonesian islands, to the Banda Sea. Located on the accompanying Sunda Arc are three-quarters of the region's volcanoes. More than 130 volcanoes have erupted—many of them repeatedly—since the close of the Ice Age. Seventy-six have erupted in just the last four centuries. Just on Java are twenty-one active volcanoes. Among the volcanoes of the Sunda Arc is Tambora, responsible for history's deadliest eruption.

Tambora is located east of Java on Sumbawa Island, about two hundred miles north of the Sunda Trench. The downward angling Indo-Australian Plate lies about a hundred miles beneath the volcano itself. The volcano has been active at least fifty thousand years, building into a towering stratovolcano some 13,000 feet above the nearby sea before its fateful eruption. By the indications of volcanic tuff on Tambora, the volcano had been quiet for five thousand years.

The first modern sign of trouble was a large explosion in 1812, followed by earthquakes and periodic lesser explosions. Many were caused by steam as groundwater seeped into contact with hot, upwelling magma. These disturbances continued until April 5, 1815.

That evening, Tambora exploded. A pillar of smoke and ash rose fifteen miles above the mountain. The explosion echoed throughout the Indonesian archipelago, from the island of Ternate, more than eight hundred miles to the northeast, to Batavia (now Jakarta), nearly eight hundred miles to the west. British soldiers garrisoned in East Java, fearing the work of insurgents, searched for the source of the explosion. Batavia Lieutenant Governor Thomas Stamford Raffles sent boats to the aid of what he thought might be a ship in distress.

Tambora rumbled and stewed for five days. Then on April 10, huge explosions drove ash twenty-eight miles high. A local headman on Sanggar about twenty-five miles east of the eruption recalled "three columns of fire rising to a great height" as lava spread over the mountain. Explosive eruptions followed for days, ultimately expelling from six to as much as thirty-six cubic miles of magma. Tambora was one of the largest volcanic eruptions in historic times, estimated at magnitude 6 to 7.

The explosions blew off as much as 4,000 feet of the mountain (now 9,348 feet in altitude) and excavated a caldera four miles across and more than 3,600 feet deep. Pyroclastic flows racing down the slopes hit the sea, producing more explosions. When the debris cooled, it formed a welded tuff up to a hundred feet thick. A half-inch of ash fell in central Java and Kalimantan, nearly six hundred miles from the eruption. Earthquakes shook Surabaya, three hundred miles away. The loudest explosions were so immense they were heard on the western coast of Sumatra, 1,500 miles to the west. Atmospheric shock shook buildings on Java and broke windows to a distance of 250 miles. Hurricanelike winds destroyed buildings. Tsunamis caused by volcanic debris or subterranean landslides flooded nearby shores. The fallen ash soaked up rain and collapsed the roofs of buildings.

"Out of a population of 12,000, in the province of Tomboro [sic], only 26 individuals survived," wrote Charles Lyell, relying on a report by Raffles. "Violent whirlwinds carried up men, horses, cattle, and whatever else came within their influence, into the air; tore up the largest trees by the roots, and covered the whole sea with floating timber. . . . The floating cinders to the westward of Sumbawa formed, on April 12th, a mass 2 feet thick, and several miles in extent, through which ships with difficulty forced their way. The darkness occasioned in the daytime by the ashes in Java was so profound, that nothing equal to it was ever witnessed in the darkest night."

As the eruption subsided and the sky cleared, bodies of people and cattle floated on the seas, buoyed along on rafts of pumice, and lay on the shores. Perhaps ten thousand people were killed by the direct effects of Tambora's fury—the pyroclastic flows, falling cinders, and tsunamis.

But worse casualties would soon follow. Ash fall—a yard deep on Sumbawa and a foot or more on nearby Lombok and Bali—buried rice fields and clogged irrigation channels. The ash, poisoning of soil, landslides, floods, and other damage to agricultural land and farming villages took a mounting toll. In all, an additional 82,000 people died of famine and disease from hunger. University of Rhode Island volcanologist Haraldur Sigurdsson puts the toll even higher—as great as 117,000. So many people died near the volcano that the local language of Tambora died out, but for forty-eight words recorded by British colonial officials shortly before the eruption.

The Tambora eruption launched fine ash and gases high into the stratosphere. As this volcanic cloud circled the earth and dispersed through the atmosphere, sulfur dioxide—tens of millions of tons of it—combined with water to form fine aerosols of sulfuric acid, which reflected incoming sunlight. As these aerosols dropped lower in the atmosphere, they provided the loci for condensation, leading to greater cloud cover, which further impeded sunlight. (Scientists learned of this phenomena from studying the similar but much smaller eruption of Mount Pinatubo in 1991.) The effect on climate was profound. Tambora's pollution lowered global temperatures by as much as 5 degrees Fahrenheit and even more at high northern latitudes. The cloud exacerbated a cooling trend that had already begun. In China and Tibet, unseasonable cold killed rice and livestock. The snow in central Europe was brown from volcanic dust. In Europe and North America, 1816 was known as "the year without summer." The weather that year and the next was cold and wet. Wheat crops failed. The bad weather followed the ruin of the Napoleonic wars and worsened food shortages, leading to urban famine, food riots, and widespread theft. In North America, New England experienced frost every month of the summer. It even snowed.

Wheat, corn, and hay crops failed. With the shortage of winter fodder, livestock died. Food prices soared, especially in large cities. Crop failures spurred thousands to quit New England for brighter fortunes in the Ohio River valley.

Lord Byron rented a villa on Lake Geneva the summer of 1816. The cold, overcast weather inspired him to write the poem "Darkness" about the aftermath of the Tambora eruption. His neighbors on the lake included English poet Percy Shelly and his bride-to-be, Mary Wollstonecraft. "It proved a wet, ungenial summer," the future Mary Shelly wrote, "and incessant rain often confined us for days to the house." To entertain themselves, they wrote ghost stories. Mary Shelly's would become *Frankenstein*.

The effect of the Tambora eruption was felt around the world. But few people understood that the change in climate was caused by a catastrophic eruption on a small island south of the equator. They didn't know because news of the event had traveled slowly or not all. All that would change with the advent of the telegraph to broadcast news of momentous events, submarine cables to carry that signal around the world, and the next monstrous eruption in Indonesia, perhaps the most famous volcanic disaster of all—Krakatau.

## KRAKATAU

Contrary to the title of the 1969 movie *Krakatoa: East of Java*, starring Maximilian Schell, Krakatau sits just west of Java. (The movie—and a popular book by Simon Winchester—both use an old English spelling of the name, which may have been the result of a telegrapher's error at the time

The 1883 explosion of Krakatau in Indonesia is one of the most infamous eruptions of all time. When it erupted in a paroxysm of fire, thunder, and smoke, it generated tsunamis and killed more than 36,000 people. *Reynold Sumayku, Shutterstock*

of the eruption. Krakatau, the Indonesian spelling, is widely used in scientific sources.) Like Tambora and the super volcano Toba, Krakatau lies near the Sunda Trench, which is responsible for the volcanic arc that forms the Indonesian archipelago. The volcano's underwater caldera is flanked by three islands: Sertung (the Dutch called it Verlaten), Panjang (Lang), and Rakata, in the Sunda Straight, not far off the busy shipping channel that runs between Java and Sumatra.

The islands have been created, modified, and collapsed by various volcanic explosions. *Pustaka Raja* (*Book of Kings*) tells the Javanese legend of an eruption of Krakatau in A.D. 416 that sundered the islands of Java and Sumatra. A possible eruption of Krakatau in A.D. 535 has been blamed for a multitude of calamities worldwide, including the fall of the Roman Empire, an outbreak of plague, the onset of the Dark Ages, the birth of Islam, Asian invaders of Europe, and the collapse of the Mayan civilization—a list that is less convincing not only for its sheer length, but because geologic evidence suggests that the volcano might never have erupted anywhere near that date.

Krakatau did erupt several times during Europe's Middle Ages and again in May 1680. After an explosive eruption in 1684, Krakatau remained quiet for two centuries, until earthquakes between 1877 and 1880, when a mighty tremor destroyed the top of the white, conical brick lighthouse on Java's Tanjung Layar, a jutting cliff at the southwestern tip of Java now in Ujung Kulon National Park. On May 10, 1883, the keeper of the light reported more tremors. Ten days later, ash and steam began to erupt from Perbuwatan, the northernmost of several vents on the island Rakata. That morning, the three-masted German frigate *Elisabeth* reported the first of several explosions, heard all the way in Batavia, a hundred miles to the east. Its captain, the first European to see and later report the eruption, described a white mushroom cloud billowing to an altitude of seven miles. By afternoon, ash began to fall on the ship and darken the sky.

On May 27, mining engineer A. L. Schuurman was dispatched aboard the *Gouverneur-Generaal Loudon* to report on the volcano. Despite the "purple, fiery glow," Schuurman ventured to the island itself in a small boat, as the larger craft kept its distance. Pumice and ash covered the beach and shore. Tree trunks and branches, though unburned, were broken and stripped of leaves, as if by a violent wind. Schuurman and other daring passengers climbed to the edge of a crater at the north end of the island. Peering in, they saw a fiery glow. As night fell, they returned to the ship.

By late June, two plumes of smoke rose from Rakata. Occasionally, several vents on the island erupted simultaneously. By August 11, a dozen vents were in eruption, ejecting steam and ash. Dutch army captain H. J. G. Ferzenaar spent two days on the island. He paddled around the eastern and northern shores and sketched a map of the island and its volcanic activity. He left the island August 12, the last person known to visit.

By August 24, the eruptions intensified. On August 26, residents of the Javan port Anyer reported the tremors of a quake. The town's telegrapher heard a huge explosion as white smoke billowed from the distant volcano. He ran to the waterfront but was forced back by a rush of water. As ash fell on the town, he ran back to the telegraph office and sent a report to Batavia that the volcano was "vomiting fire and smoke." More details followed: "Detonation increasing in loudness . . . Hails of pumice . . . Rain of coarse ash . . . first flooding . . . vessels breaking loose in harbor . . . unusual darkness . . . gathering gloom."

The *Charles Bal* was sailing northward through the straight, only ten miles from the volcano, when Captain W. J. Watson "heard above us and about the island a strange sound as of a mighty crackling fire, or the discharge of heavy artillery at one or two seconds' interval. . . . At five the roaring noise continued and was increasing; darkness spread over the sky, and a hail of pumice-stone fell on us. . . . The night was a fearful one; the blinding fall of sand and stones, the intense blackness above and around us, broken only by the incessant glare of varied kinds of lightning, and the continued explosive roars of Krakatoa made our situation a truly awful one."

The eruptions would only get worse as daylight approached. At 5:30 A.M., a tremendous explosion issued from the island. The volcano exploded again at 6:44. A hundred miles away, ash was falling on Batavia. With another big explosion at 8:20, houses in the capital shuddered.

Then, at 10:02 A.M., August 27, 1883, came the explosion the world would soon hear—either directly or by telegraph. Despite the time, the sky resembled night because of the heavy fall of ash and pumice stone. The eruption expelled more than four cubic miles of magma, a magnitude 6–6.5 explosion.

Captain Sampson of the British ship *Norham Castle* reported that "so violent are the explosions that the eardrums of over half my crew have been shattered. My last thoughts are with my dear wife. *I am convinced that the Day of Judgment has come.*" In fact, the explosion of Krakatau was one of the loudest noises in recorded history. It was heard almost immediately, of course, as a deafening thunder around the Sunda Straight. Within minutes and then hours, the sound spread throughout southern Asia, the Indian Ocean, and the South Pacific. The explosion sounded like the fire of heavy artillery 2,900 miles away on the island of Rodriguez, a dependency of Mauritius. The sound also carried to Saigon, Bangkok, Manila, Perth, the atoll Diego Garcia (due south of India, halfway to Africa), and Sri Lanka. It's estimated that the climactic eruption of Krakatau could be heard by human ears over 13 percent of the earth's surface.

The expulsion of magma actually increased the size of the islands Sertung and Panjang with accretions of tuff. But most of Rakata collapsed into the caldera formed by the depleted magma chamber, which now lay up to eight hundred feet below sea level. The salt freighter *Marie* reported "three heavy seas." At the time of the explosion, the temperature began to plummet (and would fall as much as 15 degrees Fahrenheit in four hours). Darkness closed in over the Sunda Straight that morning and lasted till the next day. Lightning that developed in the violent updraft of the eruption struck the lighthouse on Vlakke Hoek in southern Sumatra. A cloud of ash rose twenty-four miles high.

The eruption generated several powerful tsunamis, the largest, apparently, just after the 10:02 explosion, and subsequent collapse of the caldera and underwater subsidence, landslides, and volcano flank failure. Another tsunami followed more than six hours later, caused by further collapse of the caldera.

Tsunamis radiate away from their source at high speeds, scientists later learned, proportional to the square root of their depth. In the shallows of the Sunda Straight, they traveled across open water at speeds of about sixty miles per hour. At sea, the mighty submarine shocks created a wave only a few feet high. But as the wave reached shallow water near shore, it slowed, built in height, and broke on land with tremendous force.

An early-morning explosion dispatched a tsunami that raced outward across the narrow straight. On Sebisi Island at the mouth of Lampung Bay, a wave ripped away all vegetation and swept all three thousand inhabitants to their deaths.

A wave broke on the Sumatran town of Ketimbang, destroying it. At 7:45 A.M., a "giant black wall of water" from one of the earlier explosions ripped into the town of Telukbetung in the large Sumatran bay facing Krakatau. The breaking wave lifted the Dutch gunboat *Berouw*, a steamship with a crew of twenty-eight, and tossed it to the mouth of the Keoripan River. R. A. van Sandick reported four waves at high speed between 7:30 and 8:30 A.M. that "destroyed all of Telok Betong before our eyes. The light tower could be seen to tumble; the houses disappeared." The wave generated by the final explosion hit the town at 11:03 and flushed the *Berouw* two miles farther up the river valley and sixty feet above sea level.

*On Sebisi Island at the mouth of Lampung Bay, a wave ripped away all vegetation and swept all three thousand inhabitants to their deaths.*

On the Javan side of the straight, in Merak, a wave flung hundred-ton blocks of coral and limestone onto the shore and shoved a railroad locomotive nearly 150 feet off its track. The village of Anyer also was inundated.

An old Dutch pilot in Anyer who survived the onslaught of water recalled that "I turned and ran for my life. . . . In a few minutes I heard the water with a loud roar break upon the shore. Everything was engulfed. Another glance around showed the houses being swept away and the trees thrown down on every side." The rush of water swept him inland, where he clung to a palm. "The waters gradually receded and flowed back into the sea. The sight of those receding waters haunts me still. As I clung to the palm-tree . . . there floated past the dead bodies of many a friend and neighbour. Only a mere handful of the population escaped."

Within a half hour of the climactic explosion, waves higher than a hundred feet struck the Fourth Point Light near Anyer, ripping it from its foundation. Minutes later, a wave estimated at 130 feet destroyed the Tanjung Layar Light.

In all, the series of tsunamis destroyed 165 coastal towns and villages facing the Sunda Straight. Railway tracks, roads, and telegraph wires were twisted and torn down. Water swept away houses, lightly built of wood, or bamboo framing with rattan walls and palm thatch roofs. More than

36,000 people died, nearly all by tsunamis. The rest died from pyroclastic flows that shot across short stretches of ocean, borne over the water by the steam they generated.

*These atmospheric shock waves radiated out from the volcano at the speed of sound. They broke windows seven minutes later at a distance of ninety miles.*

R. A. van Sandick, author of *In Het Rijk van Vulcaan* (*In the Realm of the Volcano*), witnessed the final eruption and its immediate result at close range aboard the *Gouverneur-Generaal Loudon*: "A horrifying spectacle presented itself to our eyes; the coasts of Java, as those of Sumatra, were entirely destroyed. Everywhere the same grey and gloomy colour prevailed. The villages and trees had disappeared; we could not even see any ruins, for the waves had demolished and swallowed up the inhabitants, their homes, and their plantations."

Alexander Patrick Cameron, British consul in Batavia, a hundred miles away, would later write that "at about 11:30 A.M. at Batavia and at earlier periods of the day in the more immediate vicinity of Krakatau the sea suddenly rose, presumably owing to the subsidence of part of Krakatau and other islands or to a submarine upheaval, and a wave of considerable height advanced with great rapidity on the shore of western Java and southern Sumatra. . . . A second wave higher than the previous one followed the first at an interval of about an hour with even more serious results."

The same waves that traveled a mile a minute across the shallow waters of the Sunda Straight gained speed in the deep waters of the ocean. In the open sea, tsunamis can travel up to six hundred miles per hour and travel for thousands of miles. They measure only a yard or two high, depending on their intensity. The lateral distance from peak to trough measures miles or dozens of miles, and so the waves are virtually undetectable until they enter shallow water. Then, as the front of the wave slows, the wave grows in height. And that is exactly what happened as Krakatau generated tsunamis. The waves radiated away from the volcano, traveling for thousands of miles. In India, a wave of fourteen inches arrived at Madras. Several six-inchers washed up in Calcutta. A foot-high wave reached Karachi, Pakistan. A wave half that height reached the Yemen port of Aden twelve hours after the eruption, the distance a steamer of the day would cover in twelve days, as Tom Simkin and Richard Fiske note in *Krakatau 1883: The Volcanic Eruption and Its Effects.* A four-foot wave broke on the shore of Port Elizabeth, South Africa. The waves had enough energy to strike Europe. As Simon Winchester notes in *Krakatoa*, three-inch waves from the eruption washed into the French harbor of Socoa, 10,729 nautical miles (12,347 standard miles) from the eruption.

Shocks from the explosion traveled through air as well as water. These atmospheric shock waves radiated out from the volcano at the speed of sound. They broke windows seven minutes later at a distance of ninety miles. Barometers of Victorian-age scientists and amateur meteorologists recorded these shocks across the earth's surface. The timing of these barometric spikes indicated that the atmospheric shockwaves bounced back and forth five times during the next five days between the volcano and its antipode (a point coincidentally near Nevado del Ruiz, another killer volcano, in Colombia).

*Within three months, these pollutants had spread to high latitudes, causing sunsets so vivid fire crews were called out to extinguish imagined blazes in London, New York, and other cities.*

For days and weeks, the bodies of victims washed ashore. The Batavia *Courant* reported two weeks after the eruption that "thousands of corpses of human beings and also carcasses of animals . . . lie in knots and entangled masses impossible to unravel, and often jammed along with coconut stems among all that had served these thousands as dwellings, farming implements, and adornments for houses and compounds."

Krakatau spewed uncalculated volumes of pumice. Filled with tiny air pockets, pumice feels as light as balsa. Rafts of floating pumice drifted on the Sunda Straight and nearby seas for months following the eruption, carrying the remains of victims and interfering with shipping. A sailor on the *Samoa* steamed past Anyer "through masses of dead bodies, hundreds and hundreds of them striking the ships on both sides—groups of 50 and 100 all packed together, most of them naked." Along the eastern shore of Africa, human skulls and other bones washed ashore for nearly a year.

Ash from the eruption fell far and wide—on Singapore 500 miles to the north, on Cocos (Kneeling) Island 700 miles to the southwest, and on ships at sea as far as 3,800 miles away. But the finest particulates blew high into the atmosphere—as high as thirty miles—and traveled great distances on the wind. Even finer than these bits of dust were sulfuric-acid droplets that formed from sulfur dioxide. These high-flying aerosols reflected sunlight. The fine particulates and aerosols circled the globe at the equator in thirteen days, tingeing the sun blue and green. Within three months, these pollutants had spread to high latitudes, causing sunsets so vivid fire crews were called out to extinguish imagined blazes in London, New York, and other cities. Artists, still in the thrall of the sublime, were captivated by the intense sunsets. Between 1883 and 1886, English painter William Ashcroft painted more than 530 pastel sketches of the skies over the Thames valley.

Tennyson, in the epic "St. Telemachus," wrote, perhaps with the Krakatau eruption fresh in mind:

Had the fierce ashes of some fiery peak
Been hurl'd so high they ranged about the globe?
For day by day, thro' many a blood-red eve, . . .
The wrathful sunset glared.

The earthly cocoon of aerosols caused global temperatures to drop by more than a degree. Normal temperatures did not return for five years.

Agriculture in the region surrounding Krakatau was slow to recover. Fields had been buried in ash. Irrigation canals and rice paddies were filled in. Farmers had died by the tens of thousands. The Batavia *Courant* reported in 1910 that farm production in western Java was a third of what it had been before the eruption. By 1927, the region still produced only half of what it once had. Sumatran coastal areas lagged even further behind.

Partly because of the depravation imposed by the explosion, partly, perhaps, for psychological reasons related to the eruption, Krakatau may have been responsible for rebellion against Dutch colonial rule and the spread of Islam.

"The mullahs and teachers of religion in the pesantren, who were stirring up the people in Banten, took the opportunity given by the enormous and deep-felt impression left by the Krakatoa eruption, to expand their influence," wrote van Sandick in *Leed en Lief in Banten* (*Sorrow and Love in Banten*) in 1892. "Was it not, they said, the revenge of Allah, not only against the unbelieving dogs, but also against those Bantenese people who were serving these kafirs, these infidels? There is no doubt: the disaster at Krakatoa was a sign of God, the great omen of the holy Abdul Karim had spoken. Had he not predicted heavy earthquakes, and the end of the world?"

"On Java," writes Simon Winchester, "volcanic eruptions . . . are astral messages sent directly down to the earth, and of an importance that would be ignored only at man's peril."

Precisely because of the devastation near ground zero, Krakatau provided an unusual opportunity for naturalists and biologists contemplating the colonization of islands. What was left of Rakata was, biologically, a tabula rasa. Dutch scientist Rogier Verbeek, the first human known to have set foot on the island after the eruption, reported that six weeks afterward the rocks still radiated heat and mud still slid down the cliffs. Verbeek reported no signs of life—old or new.

Eight months later, on a French expedition to the island, biologist Edmond Cotteau reported that boulders continued to roll off the devastated slopes. The land showed no signs of mending. But he did spot a nearly microscopic spider between two rocks.

From there, it seemed, life quickly found a way to Rakata. A year after the explosion, passengers aboard passing ships spotted patches of vegetation. An expedition in June 1886 found fifteen species of forbs and shrubs, eleven ferns, and two mosses. A film of blue-green algae had begun to grow on the volcanic ash. Plant species continued to increase in number and grew ever more lush. By 1906, forests clung to the slopes. Visitors found bees, dragonflies, butterflies, kingfishers, nightjars, pigeons, swallows, bulbuls, orioles, doves, fish-eagles. Flying creatures were relatively easy to explain, since Rakata lies only forty miles from Sumatra, only slightly farther from Java. And spiders are known to travel long distances by "ballooning" on the wind beneath long strands of web. But what about the ants, snails, monitor lizards, and reticulated python? Or *Rattus rattus*, the common black rat? How had they traveled? Might they have arrived by swimming, or as stowaways on floating rafts of driftwood and pumice? Might the smaller creatures have hitchhiked on birds? By 1930, the number of species found on Rakata had reached 271 and was climbing rapidly, suggesting that the process of recolonization from the larger nearby islands was continuing apace.

A similar process played out beginning in December 1927 with the eruption of a volcano that became known as Anak Krakatau, "the son of Krakatau," from the center of the collapsed caldera. In 1930, Anak Krakatau emerged from the sea devoid of terrestrial life. By 1950, it stood five hundred feet tall and was more than a mile long. Today, it rises three times as high. Species quickly found their way to Anak Krakatau's shores and by 1990 numbered nearly 150—evidence that life is pervasive and tenacious, even in the face of unimaginable devastation.

## MOUNT PELÉE

More than 29,000 people died in the eruption of Mount Pelée on the Caribbean island of Martinique in May 8, 1902. Many of them might have been saved were a pivotal election not about to take place.

Christopher Columbus landed on Martinique, one of the Lesser Antilles, in 1502. But it was the French who began settlement of the island, in 1635. The Compagnie des Isles d'Amérique began growing cotton and tobacco and, later, sugar cane and coffee. West African slaves were shipped to Martinique beginning in the early eighteenth century. When slavery was abolished in 1848, former slaves became French citizens.

By the early twentieth century, St. Pierre was the largest, most prosperous, and most beautiful city on the island, located on an embayment along the northwestern shore of the island at the base of the volcano Mount Pelée. At the time, it was the island's principal port, a city of nearly thirty thousand known then as the Paris of the West

Indies. "St. Pierre is the quaintest and prettiest withal among West Indian cities—all stone-built and stone-flagged, with very narrow streets," wrote Lafcadio Hearn in *Two Years in the French West Indies*. "The architecture is that of the seventeenth century and reminds one of the antiquated French portion of New Orleans."

Despite its prosperity, St. Pierre was, in the spring of 1902, a restive place. A nascent socialist movement among the black and mixed-race middle class, many descended from slaves, threatened to topple the conservative island government of white planters. In a primary election held April 27, two competing socialist candidates outpolled the lone conservative; the socialists were poised to take power in the May 11 election.

The political contest, like other aspects of daily life in St. Pierre, unfolded against the backdrop of Mount Pelée, which rose impressively 4,600 feet four miles behind the city. Martinique itself was lifted from the sea on the shoulders of volcanic eruptions beginning some 16 million years ago. Mount Pelée is believed to have first erupted 200,000 years ago. Dating of the layers of tuff on the slopes of the volcano indicates there had been major eruptions in 65 B.C., A.D. 280, and A.D. 1300, with many smaller eruptions as well, including in 1792 and 1851. *Pelé* in French means "bald," in reference to the bare rock summit. The remnants of three old craters were visible. Eruptions of steam and gas had begun in 1898 and continued sporadically through the spring of 1902. Even as the primary election was held April 27, smoke and ash fell on St. Pierre. The wife of the U.S. consul wrote that "many of the people are obliged to wear wet handkerchiefs over their faces to protect them from the fumes of sulphur."

As the general election approached, earthquakes became frequent, the ash fall heavier. On May 2, the residents of St. Pierre awoke to a thin blanket of ash that lay like snow on the streets, waterfront, and buildings. A new cinder cone had appeared in the old crater dubbed L'Étang Sec, the dry pond. The next day, the rim of L'Étang Sec broke, and the resulting landslide demolished the nearby village of Le Prêcheur.

No one knew then that Martinique sits about eighty-five miles above the North American Plate, which was being driven toward the center of the earth by the overriding Caribbean Plate. The pressure and deformation of plate subduction created a ready source of molten rock slowly rising beneath the volcanic arc of the Lesser Antilles, including Mount Pelée. Magma rising within Pelée was pushing up a spine of solidified magma, a plug of rock two hundred yards across, at the rate of several yards a day. It was this spine that had emerged from L'Étang Sec. As magma escaped around the plug, the mountain began to roar.

By May 6, Pelée erupted almost continuously, ejecting red-hot bombs. The next day, explosions like heavy guns roared through the Antilles. Then, finally, at 8:02 A.M. on May 8, Pelée released four thunderous explosions from L'Étang Sec, one after another. The flaming pyroclastic flow, what French geologists call *nuée ardente*, or "glowing cloud," raced downslope to the southwest, directly toward St. Pierre.

The colonial government had downplayed the danger of the volcano. In April, St. Pierre's mayor had proclaimed the health benefits of sulfur. On May 3, the conservative *Les Colonies* had planned a public excursion to the top of Pelée to reassure the public (but cancelled the outing because of the outpouring of ash and gases). As recently as May 5, a specially appointed governor's commission of inquiry (whose only claim to scientific expertise was a single member—a high-school science teacher) had concluded "there is nothing in the activity of Pelée that warrants a departure from St. Pierre." And on May 7, Governor Louis Mouttet and his wife traveled from nearby Fort-de-France to St. Pierre.

Why were authorities acting with such complacency, even as the mountain was raining red-hot rock and ash on the populace below? The conservative government had an interest in keeping the electorate of St. Pierre in town. The conservative base of the government lived in town, and politicians were loath that wealthy citizens would evacuate before they could vote.

Mount Pelée's climactic eruption was later calculated at magnitude 4—not huge, but the circumstances couldn't have been worse. People from the countryside had taken refuge from Pelée in St. Pierre. Townspeople were at mass when the cloud of glowing ash and lava swept onto the city with greater fury and speed than an avalanche, destroying buildings and setting the town ablaze.

The pyroclastic surge swept into the harbor, sinking and capsizing ships. First officer Ellery Scott of the *Roraima* recalled, "There came a terrible downpour of fire, like hot lead, falling over the ship and followed immediately by a terrific wave which struck the ship on the port side, keeling her to starboard, flooding [the] ship, fore and aft, sweeping away both masts, funnelbacks and everything at once. . . . Shortly after, a downfall of red hot stones and mud, accompanied by total darkness, covered the ship."

Charles Thompson, assistant purser of the *Roraima* reported, "The wave of fire was on us and over us like a lightning flash. . . . I saw it strike the cable steamship *Grappler* broadside on and capsize her. From end to end she burst into flames and then sank. . . . The blast of fire from the volcano lasted only a few minutes. It shriveled and set fire to everything it touched. . . . Before the volcano burst the landings of St. Pierre were crowded with people. After the explosion not one living being was seen on land."

More than 29,000 people died, most of them instantly from burning, burial, or the inhalation of hot ash and gas. Only two townspeople were known to have survived. One had inexplicably been spared and found his way out of the burning town. The other was a prisoner in a dank, poorly ventilated subterranean cell.

On May 20, another pyroclastic flow roared over the ruins of St. Pierre and also engulfed Le Morne Rouge, a village to the northeast, on the eastern slope of Pelée. The spine that had foretold the eruption continued to grow. It

---

*Only two townspeople were known to have survived. One had inexplicably been spared and found his way out of the burning town. The other was a prisoner in a dank, poorly ventilated subterranean cell.*

---

reached a thousand feet before collapsing a year after the explosion.

The election that year was never held. The socialists were robbed of their chance for a political victory. According to Jelle Zeilinga de Boer and Donald Theodore Sanders in *Volcanoes in Human History*, "the political ascendancy of the black and mixed-race citizens of Martinique was set back for decades."

## Nevado del Ruiz

A seismograph recorded the 1985 eruption of Nevado del Ruiz volcano in Colombia. Areas known to be at risk had been mapped and the residents warned. But those precautions were not enough to prevent the death of more than 23,000 people, most killed by lahars—volcanically triggered landslides and mudflows.

Though it lies less than 5 degrees north of the equator, 17,784-foot Nevado del Ruiz is covered with snow and glaciers. A wintry storm obscured the peak November 13, preventing observers from watching a magnitude 3 explosion at 9:09 P.M. that blasted hot ash, magma, and rocks from the crater at the summit. The volume of material was small—only about 3 percent of that produced by Mount St. Helens. But as this pyroclastic flow sped across the glacier, it eroded and melted as much as 10 percent of the snow and ice on the mountain. Flows of water, ice, and volcanic debris carved channels a hundred yards wide and up to thirteen feet deep across the glaciers.

The sudden flood of water, ice, and volcanic ash surged down the mountainside, eroding gullies, turning soil to mud, and dislodging boulders and other debris. Soon these hot landslides and mudslides were racing down the slope. Some of the lahars, channeled by river valleys far down the slopes, were more than fifty yards deep and flowed downhill at speeds of nearly forty miles per hour. People who lived on high ground above the Gualí River reported seeing at least two lahars, five to fifteen minutes apart. The flood of debris shook the ground and houses and drowned out all other sounds.

Though unable to watch the eruption because of the darkness and cloud cover at the peak, scientists monitored the eruption from four portable seismographs. The sensors had been set up a few months earlier because of the volcano's well-known danger. The scientists monitoring the explosion sent warnings and details of the eruption to Colombia emergency coordinators.

Among the towns downslope of the volcano was Armero. The small city of 28,700 residents sat at the mouth of the Río Lagunillas canyon on a "debris fan"—the outwash of landslides and mudflows that had accompanied eruptions in 1595, shortly after the arrival of Spanish colonists, and then again in 1845. Each time, hundreds died. But the volcano had been quiet since, and residents had grown complacent. Which is not to say officials were unaware of the danger. A map of hazard zones had been drawn just a month before the eruption. The town of Armero, though forty-five miles from the volcano crater, clearly lay in the danger zone.

At about 11:25 P.M., the lahar approaching Armero plunged into a lake just upslope and pushed a wave of clear water ahead of it into town. Ten minutes later, a wall of mud and boulders plowed into Armero, burying homes and other buildings up to sixteen feet in debris. Just before midnight, another lahar slammed the town. Several smaller pulses nudged the town before a final large wave shortly after 1 A.M.

A flood of melt water from the Nereidas and Molinos glaciers raced down the rivers Claro and Chinchina, washing away up to three hundred houses in the city of Chinchina. By the time the lahars stopped, they had traveled more than sixty miles from the volcano. More than 23,000 people died, including three-quarters of the population of Armero.

UNZEN

Devastating lahars and tsunamis don't always follow immediately on the heels of a volcanic eruption. The eruption of Unzen in 1792 triggered Japan's worst volcanic disaster—a month after lava quit flowing.

Japan, located on the Ring of Fire around the Pacific, has about seventy-five active volcanoes. Unzen is a large complex of overlapping lava domes on the western tip of Kyushu, in far western Japan. It is located across a large bay from another landmark that would become associated with suffering and disaster more than 150 years later—the city of Nagasaki, destroyed by the second nuclear bomb dropped by U.S. forces during World War II.

The Unzen area has been volcanically active for at least 6 million years. Eruptions continued into historic times, building several large stratovolcanoes. Most recent activity has involved 4,457-foot Fugendake. It erupted in 1663, the oldest eruption recorded in historical time. It erupted again beginning in late 1791 with a series of quakes and an eruption on February 10, 1792. Large earthquakes followed. The eruption let loose a magnitude 2 explosion and continued into spring. On May 21, about a month after lava ceased to flow, nearly five hundred feet of the peak of the Mayuyama lava dome collapsed, most likely a casualty of the nearby

*The bowl-shaped scar on the mountainside remains visible more than two hundred years after the disaster.*

volcanism, and slid down the eastern slope of the mountain. The landslide roared through Shimabara City, built around feudal Shimabara Castle. The landslide destroyed much of the city and continued to slide into the sea, triggering a tsunami that raced across Ariake Bay, devastating the coast near Kumamoto (then known as Higo) with a sixty-five-foot wave. The wave rebounded to wash over the remains of Shimabara. This one-two punch destroyed 3,300 homes and killed nearly 15,000 people. The bowl-shaped scar on the mountainside remains visible more than two hundred years after the disaster.

The 1792 eruption would remain Japan's greatest volcanic disaster—but it would not be Unzen's final eruption. In February 1991, the volcano stirred again. By May, the volcano produced a new dome. During the next four years, the growing dome repeatedly collapsed, discharging ash flows downhill at speeds up to 120 miles per hour. During the summer of 1991, some 12,000 residents were evacuated. A pyroclastic flow on June 3 sped three miles downslope and killed forty-three people who had entered a known hazard zone. During the next three years, the mountain produced about ten thousand pyroclastic flows and land-slides (especially during the rainy season), which destroyed some two thousand buildings. Japanese public safety offi-cials built a series of dikes and sediment basins on the eroded slope to channel and contain landslides and lahars.

## Laki

Iceland sits atop the Mid-Atlantic Ridge, a zigzag seam of the Atlantic where the sea floor pulls apart and magma surges up at a ferocious rate to fill the gap. Iceland is part of the ridge. In fact, the island itself is being pulled apart. The eastern half, part of the Eurasian Plate, is sailing east, fol-lowing Europe at the rate of about a centimeter a year. The western half is part of the North American Plate, navigating westward at a similar rate. And through the middle of the island, upwelling magma pours into the split. The result is a volcanic field that runs north-south through the country and erupts frequently—sometimes with disastrous results.

Iceland's position is doubly volatile. Not only does it stand astride the rift between two plates, it also sits above an active hot spot created by a plume of magma rising from the upper mantle. The heat and force of the hot spot has uplifted the crust, including Iceland. Scientists disagree whether the rising plume is forcing the plates apart or if the separating plates allow the plume to rise. Either way, the plates are spreading, the magma is rising.

Iceland was formed by upwelling magma. Thomas Carlyle captures the dichotomy of fire and ice in Iceland's formation by volcanism: "Iceland—burst up, the geologists say, by fire from the bottom of the sea; a wild land of barrenness and lava; swallowed many months of every year in black tempests, yet with a wild gleaming beauty in summertime; towering up there, stern and grim, in the North

*Oh, what terror it was to contemplate such signs of fury, such divine manifestations! . . . it was a suitable time to talk to God!*

Ocean; with its snow-yokuls, roaring geysers, sulphur pools, and horrid volcanic chasms, like the waste chaotic battle-field of Frost and Fire."

The setting on Iceland is more complicated that a single rift. There are two splits, side by side. The rift to the east, running from northeastern Iceland down through the Vestmann Islands, is the zone where most of Iceland's active volcanoes lie.

Since settlers arrived from Norway in A.D. 874, more than thirty volcanoes have sprung to action. Rhyolitic crust often mixes with the upwelling basalt to increase the viscosity of the magma. Since settlement, twenty-seven eruptions have exploded with magnitudes ranging from 4 to 5. In other words, big-time explosions. These outbursts have played to Iceland's pre-Christian visions of their gods battling at the end of the world, as recounted in the prophetic *Völuspá*:

> The sun begins to be dark; the continent falls fainting
> into the Ocean;
> They disappear from the sky, the brilliant stars;

The smoke eddies around the destroying fire of the
world;
The gigantic flame against heaven itself.

Hekla was the volcano most likely to inspire such visions, erupting an average of every forty years. Its stirrings have been traced to an explosion in A.D. 1104, the largest eruption in historic Iceland. "This mountain," wrote French cleric Herbert of Clairvaux, "all burning and belching flame, stands in a perpetual blaze, which spreads over the mountain."

Yet Iceland's deadliest volcano would erupt a bit to the northeast, near Vatnajökull Glacier, where a deep fissure had created a chain of at least 140 volcanic vents and spatter cones, including an old volcano known as Laki Mountain. Beginning June 8, 1783, after earthquakes had rocked the region for three weeks, a fissure up to six miles deep rent the mountain and began to spill fantastic volumes of lava and gas in a series of explosions up to magnitude 4. Flowing lava filled the Skaftá River gorge and then spilled over to the Hverfisfljót River. The lava pushed down both valleys to the coastal plain. There it began to spread, burying churches and farms. "Big blocks of stone and sods of turf were thrown into the air to an unspeakable height, from time to time accompanied by great bangs, flashes, jets of sand and light or dense smoke," wrote local pastor Jón Steingrímsson. "Oh, what terror it was to contemplate such signs of fury, such divine manifestations! . . . it was a suitable time to talk to God!" The eruptions lasted until the following year, ejecting about three cubic miles of lava covering more than two hundred square miles, the greatest on-land outpouring of lava observed in historic times.

Yet it was not the lava itself that did the damage. Tremendous volumes of ash, water vapor, carbon dioxide, sulfur dioxide, hydrogen chloride, and fluorine rose into the stratosphere. The estimated 50 million tons of sulfur dioxide formed sulfuric-acid aerosols, diffusing and reflecting sunlight. The pollutants produced haze and acid rain from Iceland southwestward into northern and central Europe, bringing cold to a region already in the throes of the Little Ice Age. "Sulphurous rain caused such unwholesomeness in the air and in the earth that the grass became yellow and pink and withered down to the roots," Steingrímsson wrote. Farmers watched their livestock grow lame from fluorosis, a crippling disease that produces lesions on gums, teeth, and bones. Half of Iceland's cattle and most of its sheep and horses died in the year following the eruption. Settlers referred to a "blue haze famine." Cold, stormy seas produced fewer cod than normal. Nearly ten thousand people, about 20 percent of Iceland's population, died from starvation or disease. Elsewhere in Europe, the atmospheric changes caused widespread crop failures.

During the Icelandic eruptions, Ben Franklin represented the United States in the court of Louis XVI. Even in France, he noticed the stinging haze and preternatural cold. Franklin was the first to speculate in print that volcanoes may be to blame for widespread climatic change. "The cause of this universal fog is not yet ascertained," Franklin wrote in a paper, "Meteorological Imaginations and Conjectures," read before the Literary and Philosophical Society of Manchester in 1784. "During several of the summer months of the year 1783, when the effect of the sun's rays to heat the earth in these northern regions should have been greatest, there existed a constant fog over all Europe, and great part of North America. This fog was of a permanent nature; it was dry, and the rays of the sun seemed to have little effect towards dissipating it, as they easily do a moist fog, arising from water. They were indeed rendered so faint in passing through it, that when collected in the focus of a burning glass, they would scarce kindle brown paper. . . . Whether it was adventitious [coming from space] to this earth, . . . or whether it was the vast quantity of smoke, long continuing to issue during the summer from Hecla, in Iceland, . . . which smoke might be spread by various winds over the northern part of the world, is yet uncertain."

There is some evidence the effects of Icelandic volcanism spread throughout the northern hemisphere. The winter of 1783–84 was so bitter in North America that the Mississippi froze at New Orleans.

Magma has continued to pour out over Iceland. In November 1963, smoke, ash, and steam bubbled to the surface of a disturbed sea in the Vestmann archipelago. On November 15, an island rose above sea level. The new feature, which now stands 570 feet above the water, was named Surtsey for the Norse giant Surter, who carried a burning sword.

In 1973, a volcanic eruption on Heimaey Island buried homes and commercial buildings in the town of Vestmannaeyjar and threatened to fill in the island's harbor, Iceland's largest fishing port. Heimaey's residents were evacuated, but the fight was on to save the harbor as lava advanced three to nine yards a day. Large pumps and the dredging ship *Sandey* were brought to the island to hose down the lava with a ton of seawater a second. Workers managed to cool the advancing edge of the mass and form a dam that held back additional volumes of lava. The eruption slowed and stopped. The harbor was saved. In fact, with a newly formed wall of basalt to narrow the opening, the harbor was more secure than ever.

SANTORINI

Though there is no body count to prove it, there's every reason to believe the explosion of Santorini Island in the Aegean Sea in about 1600–1650 B.C. was one of the great natural catastrophes in human history. The explosion, perhaps as great as magnitude 7, was one of the most powerful in the last ten thousand years and bordered on super volcano strength. The island, and neighboring islands, clearly had been inhabited. In fact, there is some good reason to believe that the eruption may have lead to the downfall of the Minoan civilization of Crete and even gave rise to the legend of the city that fell into the sea—Atlantis.

Santorini, officially known as Thera, is a circular cluster of volcanic islands poised southeast of the Greek mainland, southwest of Turkey, and north of the island of Crete. Thera is also the name of the largest island, which brackets a northern caldera and a southern caldera. These calderas are among several volcanoes in the South Aegean, or Hellenic, Volcanic Arc, the result of the subduction of the African Plate beneath the Aegean Subplate. The Santorini islands have a long, violent history, with evidence of hundreds of eruptions during the last 2 million years. During the past 200,000 years, eruptions have built up large shield volcanoes, only to destroy them in periodic explosions.

The oldest signs of human settlement on the islands go back as far as six thousand years. By about 1650 B.C., Thera had become one of the Aegean's most important ports. Recovered trade goods originated in Anatolia, Cyprus, Syria, Egypt, and Greece. Thera was an outpost of the Minoan civilization of Crete. Their art of dolphins, fishermen, and boats is preserved on the walls and pottery buried beneath the debris of this Bronze Age eruption. Pipes carried running water through the city. Twin pipes suggest residents even enjoyed running hot water, piped from the steaming hot springs around them.

Little is known about the lead-up to the catastrophic eruption of Santorini. In *Volcanoes in Human History*, de Boer and Sanders describe an eruption in four stages: an explosion of ash and pumice high into the atmosphere. Ash and coarser debris covered the surrounding islands, leaving them uninhabitable. Then followed quiescence, perhaps up to twenty years. Some buildings showed signs of repair. Then seawater infiltrated the crater, producing an explosion and lahars that deposited mud and ash up to forty feet thick. Later, as seawater poured into the magma chamber, Santorini really blew, producing an explosion as powerful as magnitude 7. The explosion ejected from seven to perhaps as much as twenty-four cubic miles of magma and created a large caldera. The explosion propelled a column of ash far into the atmosphere. Blankets of ash fell on islands throughout the Aegean. The eruption would have been heard throughout the ancient Mediterranean.

What happened to the island residents? Excavations in Akrotiri, a city on the south of Thera, have revealed walls, homes, streets, and plazas but have produced no bodies at all. Even moveable objects appear to have been removed. Had a long ominous prelude of earthquakes and lesser eruptions given the populace plenty of time—months or even years—to evacuate?

Even if Minoans had abandoned Santorini, they would not have been spared the volcano's deadly effects. The climactic eruption produced a tsunami that broke, perhaps higher than a hundred feet, on the northern coast of Crete, just forty-five miles away. The wave would have destroyed the Minoan fleet and countless seaside towns.

The ejection of so much fine ash and sulfur into the atmosphere would probably have cooled the climate, on the order of the year without summer following the similar-sized Tambora eruption of 1815. Greenland ice cores show a rise in acidity. Chinese records from about this time tell of yellow fog, dim sunlight, summer frost, and widespread famine. The change in weather slowed the growth of trees in Turkey and as far away as Sweden.

Crete lost its population, and Minoan culture faded during the Bronze Age. Greek archeologist Spyridon Marinatos, who pioneered the excavation on Thera, suggested in a 1939 article in the British journal *Antiquity* that the explosion brought an end to Minoan civilization.

The destruction of Santorini provides a prototype for the Atlantis legend of a city sunk to the bottom of the sea

*Chinese records from about this time tell of yellow fog, dim sunlight, summer frost, and widespread famine.*

by angry gods—a story recalled by Plato in two of his dialogues, *Timæus* and *Critias*: "The Gods' anger against Atlantis was so strong, that they destroyed it in a single day and night, by earthquakes, and sunk it into the sea, leaving only a mass of mud behind."

The aftereffects of the Santorini explosion may have been more profound, even, than the creation of a lost city. As Minoan culture weakened, Mycenaean Greeks conquered Crete, explain de Boer and Sanders in *Volcanoes in Human History*. "Thus the Bronze Age eruption of Thera and the resulting Minoan diaspora appear to have been major factors in the rise of Mycenaean Greece, the cultural ancestor of classical Greece [which] . . . gave rise to the values and philosophies that underlie the culture we call Western civilization today."

SANTA MARÍA

The Santa María eruption in 1902, a magnitude 6 blast that was the second most explosive eruption of the century, killed as many as six thousand in southwestern Guatemala.

The stratovolcano, now 12,360 feet, is one of several that rise above the Pacific coastal plain of Guatemala and Honduras, near the junction of the North American, Cocos, and Caribbean plates. People have lived near its base for thousands of years. An Olmec archeological site lies just south of the mountain. At the time of the Spanish conquest of Central America, the city of Quetzaltenango was the center of the Quiché kingdom Xelajú. Quiché Indians continued to live in the city after the conquest, and in the centuries since, Quetzaltenango became a center for trade between the coast and mountains.

In 1902, Santa María had been dormant for hundreds, perhaps thousands, of years. But in January began a fit of earthquakes that continued for months. In April, a magnitude 8.3 quake killed 1,400 in Quetzaltenango, about twelve miles from Santa María. Clearly, something was brewing. Perhaps to allay fears, or perhaps to prevent concern over a restless volcano from interfering with state Feast of Minerva festivities, Guatemalan dictator Manuel Estrada Cabrera declared on October 24 that all the country's volcanoes were calm. But at least one wasn't. Just that day, Santa María began to spew ash and volcanic debris on the city below.

The next day, the volcano let loose with an explosion that unleashed a column of ash sixteen miles high and blew out more than a cubic mile of pyroclastic debris from the southwestern slope of the mountain. The resulting landslide, pyroclastic flows, and lahars swept away more than a hundred villages surrounding Quetzaltenango.

The disaster would not be Santa María's last. In 1922, a volcanic dome began to grow from the crater of the volcano. It soon became known as Santiaguito. In 1929, the dome collapsed. Pyroclastic flows raced downslope for six miles, killing as many as five thousand in villages and plantations below. Santiaguito remains an active threat to the estimated 300,000 residents of present-day Quetzaltenango. "The danger has always been there," said a friend who grew up in Quetzaltenango, and whose family remains there still. "It is the risk they are born with and the risk they will die with."

## KELUT

The 5,679-foot stratovolcano Kelut rises above farmland and populous villages in central Java. People have always lived near the volcano. From the eleventh to thirteenth centuries, Kelut sat amid the most powerful kingdom of the Indonesian islands. From the fourteenth to sixteenth centuries, Candi Penataran, located at the foot of the volcano, became the largest and most important Hindu temple complex in the region.

The people who inhabited the fertile plain lived a perilous existence since Kelut has been one of the most active volcanoes in one of the most volcanically active areas of the world. Kelut has erupted more than two dozen times since A.D. 1000, fifteen times in the last two hundred years. Its eruptions have been short and explosive.

Kelut posed the usual hazards—pyroclastic flows, landslides, and poisonous gases. But it has posed a unique danger as well: Its large crater would fill with water. The eruptions would discharge crater lake water down the side of the mountain. The flood would loosen debris, triggering landslides and deadly lahars that would bury villages beneath speeding boulders, mud, and water.

Kelut's lahars, pyroclastic flows, and surges have killed more than 15,000 people since A.D. 1500. An eruption in 1586 was particularly large, an estimated magnitude 5. The resulting lahar killed ten thousand. The volcano has been dangerous, even when dormant. In 1875, when the volume of the crater lake had reached 78 million cubic meters, a heavy rain caused the crater rim to break. Half the lake cascaded down the southwestern slope, creating a mudslide eight miles long that destroyed the villages of Sregat and Blitar. Another eruption in 1901 triggered enough concern that the Dutch administration built a dam along the river Badak to shield Blitar from lahars.

But the preparation was inadequate. A daylong magnitude 4 blast that began May 19, 1919, produced a lahar that destroyed the Badak dam. Lahars from the eruption traveled twenty-three miles in less than an hour, burying or flooding nearly forty thousand acres of farmland, destroying a hundred villages, and killing 5,160.

Right after the disaster, Indonesia created the Volcanological Survey to tunnel through the Kelut crater wall to drain the lake to a safe level. The lake was dry when excavation began from both sides of the crater wall. But ground temperatures of 115 degrees Fahrenheit slowed the work. The project had not been completed when, in 1923, the lake rose and flooded a tunnel with mud, killing five workers. Plans were changed: Engineers designed a series of tunnels at various levels. The job was completed in 1926 and succeeded in keeping the lake at a reduced—and safer—level. Even so, an eruption in 1951 enlarged the crater and damaged tunnels. After two hundred died in a 1966 eruption, workers dug a new and deeper tunnel.

A large population lives in fear of another eruption. Kendiri, just sixteen miles from the Kelut crater, now numbers 1.3 million. And the volcanic plain is packed with farms raising rice, sugarcane, pineapple, tobacco, and other crops.

## GALUNGGUNG

Galunggung, a 7,111-foot peak in western Java, has been one of the country's most active stratovolcanoes. Like Kelut, it presents the threat of deadly mudslides. Unlike almost any other volcano in history, it nearly brought down two commercial jetliners.

---

*As a British Airways Boeing 747-200 carrying 247 passengers and a crew of 16 flew at an altitude of 37,600 feet through ash nearly a hundred miles downwind of the crater, the plane's four engines quit.*

---

Galunggung provides plenty of evidence of long-ago activity. At its base are the Ten Thousand Hills of Tasik-malaya, a plain of avalanche-debris hummocks formed during the prehistoric collapse of the volcano caldera. The first historic eruption occurred in 1822—and it was big, magnitude 5. Pyroclastic flows sped along more than six miles from the crater. Mudslides killed 4,011 and destroyed 114 villages.

Since then, the volcano has erupted four times, most recently in 1984. A magnitude 4 explosion in 1982 killed nearly seventy, mostly indirectly through traffic accidents and hardship and trauma to the elderly. But the most spectacular and unusual threat of the eruption was the cloud of volcanic ash. The fine volcanic ash melts in a hot jet engine, forming a thin coat of glass on the fuel injectors. Accretions of ash adhering to the glass build up, stalling the turbines. As a British Airways Boeing 747-200 carrying 247 passengers and a crew of 16 flew at an altitude of 37,600 feet through ash nearly a hundred miles downwind of the crater, the plane's four engines quit. The aircraft plunged more than 24,000 feet. The crew managed to restart the craft only 13,000 feet above the ground. The next month, a Singapore Airlines plane with 230 passengers flew through the ash cloud and lost three out of four engines. After a dive of 7,800 feet, the crew restarted one engine and, like the British plane, made an emergency landing in Jakarta.

Vesuvius

The Italian volcano Vesuvius, the only active volcano on the European mainland, has erupted more than fifty times since the Bronze Age. Some eruptions were deadly. An explosion in 1631 is blamed for about 3,500 deaths, and some put the toll as high as 18,000. But we know the name Vesuvius for its eruption in A.D. 79. That eruption preserved a moment in time. And for that reason, it has become one of the most famous and most studied volcanic events in history

Rising from the plain of Campania, between the Apennine Mountains and the Tyrrhenian Sea, Vesuvius provides a striking backdrop to the Bay of Naples. Two old Oscan villages on Vesuvius's slopes grew into the towns of Pompeii and Herculaneum. Both rebelled against Roman rule. The revolt was put down, and the Romans made Pompeii a colony for veterans of Rome's armies. Pompeii flourished

The violent eruption of Mount Vesuvius in A.D. 79 buried the city of Pompeii, Italy, in fifteen feet of ash. Excavations have unearthed, in the words of Mark Twain, "long rows of solidly-built brick houses (roofless) just as they stood eighteen hundred years ago." *John Lumb, Shutterstock*

and became a center of trade, with a population of twenty thousand and many luxurious homes. In an amphitheater, the city's entire population could watch contests between gladiators. Herculaneum, meanwhile, was much smaller—perhaps five thousand—a commercial fishing village where some of Rome's wealthiest built villas.

The towns sat on a magma chamber more than a mile high and nearly a mile across, buried about three miles below the surface. The magma welled up from a complex subduction zone. The area had been volcanically active for perhaps 300,000 years. The Mount Vesuvius stratovolcano grew inside an older caldera. Eight major explosions occurred in the last 17,000 years. Eruptions in 5960 B.C. and 3580 B.C. were powerful, perhaps magnitude 5. Recent evidence of a large eruption about 1780 B.C. includes thousands of footprints in volcanic ash, all running from the crater. Charred skeletons show that many didn't escape. There was another historic eruption in 1550 B.C. and perhaps another in 217 B.C., when, Roman poet Silius Italicus wrote much later, a "fiery crest, throwing rocks up to the clouds, reached to the trembling stars." But by A.D. 79, the volcano had been quiet long enough that citizens of thriving Pompeii risked living at the volcano's base.

An earthquake in A.D. 62 forewarned of the disaster soon to follow. The tremors cracked pavement and walls, and collapsed roofs. But life passed for normal. On August 24, 79, people were shopping at the markets, lounging at the public baths, drinking in bars, and resting and working at home when Vesuvius erupted with a magnitude 6 explosion that shot a column of ash twenty miles into the atmosphere.

Pliny the Younger would describe it as "being like an umbrella pine, for it rose to a great height on a sort of trunk and then split off into branches." (It is for Pliny that explosive eruptions like Vesuvius are today called *plinian*.)

A pyroclastic surge, the advance wall of hot gases and glowing small particles, blew into Herculaneum at speeds up to two hundred miles per hour, ripping off roofs, blowing down buildings, and killing any living thing. Heavier pyroclastic flows followed, setting everything but stone or metal aflame. The flows buried the little town more than sixty feet deep.

The volcano's first and second pyroclastic flows never reached the city of Pompeii. The third damaged northern sections of town. On the second day of the eruption, a fourth flow buried Pompeii to depths of more than fifteen feet. About a cubic mile of magma and ash covered nearly two hundred square miles. About four days after the eruption had begun, it ended. And, except for the roofs of some buildings in Pompeii, the two towns would disappear for nearly two thousand years.

There is some evidence that people—either residents or thieves—had tried to excavate parts of the buried towns. But the effort was abandoned. It was not until about 1595 that workmen digging a canal from Sarno to a nearby village rediscovered the forgotten settlements. Centuries of pillaging followed. In 1689, well-diggers discovered a stone with the Roman inscription *Pompeii*. At about that time, Giuseppe Macrini tunneled into an area known as la Città, "the city." Soon after, he described evidence of houses and walls in a book advancing this theory that the site was ancient Pompeii. Plundering continued. It wasn't until the unification of Italy in the mid 1800s that systematic excavation of the two towns began.

Only farm fields and a few yards of compacted ash and other loose debris covered Pompeii, so the remains were easy for archeologists to reach. Herculaneum, on the other hand, had been buried by deep lava flows. Moreover, the town of Ercolano had been built on top, so archeologists could reach the old town only by tunnels. Nonetheless, during the next century, much of both cities would be excavated.

Pompeii's remains revealed much about how people both lived and died. Victims who had taken shelter in buildings were asphyxiated or crushed as roofs collapsed.

*Workers discovered many of the molds of people buried in ash, their final moments forever preserved.*

Bodies were covered by loose ash and debris. Skeletons were often discovered in crouching poses. Some were adorned with jewelry. Coins that had dropped from rotting purses or pockets often lay next to bones. People caught outside were buried in deep ash. With time, the ash consolidated, preserving the shape of bodies within even as they decayed. Workers discovered many of the molds of people buried in ash, their final moments forever preserved. Plaster of Paris poured into the molds reproduced the bodies, even down to details of clothing and facial expressions.

No similar molds existed in Herculaneum because of the heavy pyroclastic flows. Archeologists did discover more than two hundred skeletons in boathouses along the

A plaster of Paris mold reveals the final moments of a Vesuvius victim buried by heavy ash. *John Lumb, Shutterstock*

Herculaneum waterfront. Apparently, victims had packed into the shelters and were killed instantly by a pyroclastic surge. According to a University of Naples team, the victims died so quickly they didn't have time to raise their hands to ward off the heat.

Some residents, especially in Pompeii, probably escaped in the hours before the final pyroclastic flows. But once the hot ash and lava entered the city, few survived. The death toll from Vesuvius has been commonly accepted at 3,360, based largely on discovered remains. But more recent estimates, based on the discovery that pyroclastic flows reached much farther than first believed, have pushed the death toll in the cities and surrounding countryside much higher—to perhaps 16,000.

But it is not for the body count that we remember Vesuvius. Rather, it is for the irony that by destroying Pompeii and Herculaneum, Vesuvius preserved the towns for another day. Though excavation was still in its early stages when American author Mark Twain toured Pompeii in 1867, he was moved by the sight: "There stands the long rows of solidly-built brick houses (roofless) just as they stood eighteen hundred years ago, hot with the flaming sun; and there lie their floors, clean-swept, and not a bright fragment tarnished or wanting of the labored mosaics that pictured them with their beasts, and birds, and flowers, which we copy in perishable carpets today; . . . and there are the narrow streets and narrower sidewalks, paved with the flags of good hard lava, the one deeply rutted with the chariot-wheels, and the other with the passing feet of the Pompeiians of the by-gone centuries."

Mount Vesuvius, an active volcano to this day, continues to threaten a million people who live and work within four miles of the crater. *Danilo Ascione, Shutterstock*

Today, a million people live and work within the four-mile radius of the crater—the area that pyroclastic flows and debris slides could destroy in the first fifteen minutes of a medium- to large-scale eruption. Eventually, 3 million people could be at risk. Professor Franco Barberi of Rome University told the 32nd International Geological Congress in Florence in 2004 that Vesuvius is the world's most dangerous volcano.

Humans, unfortunately, are much practiced in the art of denial, whether the warnings of experts or the rumblings from a mountain—even the ominous signs issuing from one of the largest volcanoes ever to erupt on the face of the earth.

# CHAPTER 9

# THE NEXT BIG BLAST

"Civilization exists by geologic consent, subject to change without notice."
— Will Durant, philosopher and historian

If volcanoes of middling power can kill many, what might happen if Yellowstone were to let loose with another caldera-forming blast?

Since modern humans have never experienced the eruption of a super volcano firsthand, the answers are necessarily speculative. From the record of Yellowstone's past eruptions, traced in the rocks of the park and the Snake River Plain, scientists can infer that the destruction would be vast, covering thousands of square miles. And from what we know from the effects of lesser volcanoes, scientists can also infer that the human toll would be unfathomable. Short of launching a full-fledged investigation by a blue-ribbon committee of specialists, it's tough to anticipate the myriad difficulties caused by the volumes of magma, ash, and gases unleashed in such an explosion. We'd expect the tragedies of large volcanoes in the past—pyroclastic flows, floods, lahars, poisoning, choking ash fall—multiplied by a hundred. Or a thousand. It would be a scene from the apocalypse. It might indeed *be* the apocalypse.

Assume the Yellowstone volcano gets its preliminaries out of the way and explodes as it did 2.1 million years ago. Within hours or, at most, days, the Yellowstone Plateau bursts at the seams with explosions of magma more powerful than the largest manmade nuclear blasts. Mountains vaporize. Billowing clouds of water vapor, ash, and gases vault to the stratosphere, lit up by lightning storms. The entire Yellowstone Plateau falls into the emptying magma chamber, forming the caldera and launching even more explosions.

Pyroclastic flows 1,000 degrees Fahrenheit—"incandescent hurricanes" as they've been called, a "blanket of utter devastation"—race outward from the volcano, down the valleys of the Madison, Gallatin, and Yellowstone rivers, southward through the valley of the Snake, fiery hands reaching as far as eighty miles. Red-hot rock cools as welded tuff from 500 to 2,500 feet thick over 6,000 square miles. Lahars charge down river valleys.

Yellowstone National Park is destroyed, the forest leveled, every distinctive feature obliterated, every living thing vaporized, incinerated, blown to bits, buried in welded ash and mud, or asphyxiated by poisonous gases. That's also the prognosis for the towns surrounding the park, such as West Yellowstone, Gardiner, and Cooke City.

"You'd be dead," said Michael Rampino.

Rampino is associate professor of Earth and Environmental Sciences at New York University. He has studied mass extinctions and the effects of huge asteroid impacts on

The magnitude 5 eruption of Mount St. Helens in 1980 blew out the side of the mountain, destroyed forests for miles around, caused $3 billion worth of damage, and killed fifty-seven people. The U.S. Geological Survey described it as the "worst volcanic disaster in the recorded history of the United States." A super eruption at Yellowstone would be a thousand times more powerful. *Austin Post, Weatherstock*

life on Earth. In his studies, he has realized a super eruption would produce similar results. He has researched the effects of eruptions on climate change and the impact of volcanic winter on human life and civilization. I asked him to walk me though what might happen if Yellowstone unleashed another caldera-forming explosion.

Even more distant towns of some size, such as Cody, would be in danger, he said. "You'd be in serious trouble. Pyroclastic flows would probably go that far, and ash fall would be very, very heavy. So it's quite unlikely you'd survive.

"I would say within a hundred miles of the volcano would not be a good area. Think of Mount St. Helens—a tiny little volcano."

And think of the magnitude of the Mount St. Helens eruption May 18, 1980. Charles McNerney and John Smart, along with several other people, stood on an overlook eight miles from the volcano when it erupted. The two men watched as a dark cloud, the pyroclastic surge, sped toward them and began to climb the ridge. Unlike the others, who hesitated (and did not survive), McNerney and Smart jumped in their car and barely outran the cloud at speeds over eighty miles per hour. The blast swept over loggers working twelve miles away and out of sight of the volcano. Toward the north, the surge traveled some twenty miles, at speeds of more than three hundred miles per hour. Victims at a distance of several miles were killed by falling trees, burned, and asphyxiated. The damage in Washington State alone was calculated at nearly $3 billion.

The range of a Yellowstone explosion—potentially a thousand times more powerful—would be much greater.

Depending on the amount of warning, tens of thousands might die instantly, even though the area surrounding the park is mostly rangeland, forest, and scattered small towns and settlements. But even beyond the extent of pyroclastic flows, the Yellowstone blast would have far-reaching effects. Even global ones. And there's not much we could do to escape.

The clouds of tephra, ash, and volcanic gases blown in all directions begin to fall to earth, the heaviest and coarsest dropping out first. Finer particles, launched high into the atmosphere, catch the prevailing westerlies. The very finest hitch a ride in the jet stream. Towns and cities near the volcano—Bozeman, Billings, Idaho Falls, and Casper—will have to dig out, as if after a blizzard. Even Denver and Salt Lake City will have to call out the plows to remove a foot of ash. Heavy ash falls across the Great Plains and quite possibly Canada's Prairie Provinces. Winds rake the ash into drifts, and fill swales to thirty feet deep.

---

*Airplanes won't be flying. Internal combustion engines will be screwed up. Trucks won't be going.*

---

Radio and phone communications black out as electrically charged ash disrupts radio waves, and air-cooling systems for electronic switching gear are shut down or clogged. Heavy ash shuts down large power generators. Ash absorbs rain, causing electrical insulator flashovers. Ash on roofs soaks up water, collapsing buildings.

Residents of the West face a vexing problem. They survive the ash fall (except, perhaps, those with respiratory diseases). But now what? Most are not able to leave. For days or weeks, cars, buses, and planes are stranded because the drifts of ash block roads. Wet ash slicks up the pavement. Airborne ash drives visibility to near-zero. Ash clogs air filters, fouls oil, and abrades metal parts, stalling and destroying engines in record time. In the days following the blast, food, medical supplies, and other goods arrive in areas of heavy ash fall only by the most strenuous efforts. "Airplanes won't be flying. Internal combustion engines will be screwed up. Trucks won't be going. So yes," said Rampino, "you'll have a big problem with transportation." The best hope for shipping and travel might be the railroads.

---

*As the average temperature drops several degrees, the planet plunges into a decade-long volcanic winter.*

---

"How do you get food, how do you get supplies, how do you get in and out, even after the eruption?" asked Rampino. When asked how many will be killed, he responds, "It depends on what you mean by killed—killed right away or starved to death?"

The specter of starvation extends over an area much larger than the West. The Great Plains wheat fields, the global breadbasket, source of up to half the world's cereal grains, are buried by ash for a season—or several. Where soil is covered by six inches or more of ash, crops and pasture grasses die; so also do the microorganisms that make soil productive. Harvests are destroyed. Fields lie fallow for years and will be reclaimed only at great expense, by tillage with high inputs of fertilizer. The cost of restoring fields might kill Great Plains agriculture, a marginal business under any circumstances. According to U.S. Geological Survey disaster planners, where ash is deeper than a foot, land will be untillable "for many generations."

Like the rhinos at Ashfall, domestic livestock die quickly from starvation, respiratory distress, or more slowly by fluorosis from the soluble fluorine on pasture grasses. Until ash consolidates, it absorbs water, so little is available to livestock. Animals are slaughtered, but with limited transportation, what can be done with the sudden surplus of meat?

In the Midwest, where an inch to several inches of ash has fallen, major rivers fill with sediment. Power plants close to protect turbines and other machinery from the abrasive ash. With the increased turbidity in tap water, officials tell residents to boil water to prevent disease. Cities ration water use. Ash blocks sewers and pipes that carry wastewater. As much as eight hundred times the normal load of sediment chokes the Mississippi. It's the Big Muddy in spades. Channels clogged with ash flood from the slightest rain. The sediment-filled river is useless for barge transport until the ash is dredged or gradually flushed downstream. The transportation route for much of the nation's—and the world's—grain closes down. "The whole world relies on us," Rampino said. "If you wipe that out for a growing season or two, you have worldwide famine problems."

While ash, even fine ash, gradually settles out of the atmosphere, carbon dioxide, hydrogen chloride, and hydrogen fluoride stay aloft much longer. Sulfur dioxide converts to sulfuric acid, which condenses to form sulfate aerosols. A vast cloud of these aerosols in the middle and lower stratosphere encircle the globe within weeks of the eruption. These clouds both reflect and absorb the sun's radiation, causing the earth's upper atmosphere to warm and the lower atmosphere to cool. As the average temperature drops several degrees, the planet plunges into a decade-long volcanic winter. As if ash-covered grain fields aren't enough, low temperatures cause poor harvests. The result: Even many regions of the world that are normally self-sufficient in food will face famine.

How many might die? I asked Rampino. "People who have talked about it have talked about a billion-plus people," he said. "One-seventh of the world's population." Even if Rampino were off by a factor of 100, there has never been a natural disaster that comes close.

The physical effects and body counts, though speculative, are rather straightforward to consider. Much more difficult to calculate are the political, religious, and strategic effects of a super eruption. Imagine, for example, that Yellowstone were to unleash a magnitude 8 explosion in the next few years, when the world is divided by religious fervor. If the United States is crippled by a natural disaster of such magnitude, its enemies will certainly perceive it as vulnerable. It's hard to imagine that the country, smothered by ash, would stand ready to fulfill geopolitical commitments to allies. For that matter, what would be the various religious interpretations of such an event—as a cause for ever-greater fundamentalism? And countries and regions already threatened by rising seas and stressed by climate change—what resilience will they have in the aftermath of a super eruption? Clearly the United States would be in no position to deliver aid. And if Yellowstone were to erupt a century from now, or a millennium, who could begin to predict the geopolitical forces that might be set in motion?

A 2005 report on the devastating effects of super volcano eruptions by the Geological Society of London bleakly concluded, "The effects could be sufficiently severe to threaten the fabric of civilisation."

— . —

As devastating as a Yellowstone super eruption might be, what are the chances it will ever happen?

Very good, actually. Scientists say another eruption is all but inevitable. But the chance of it happening anytime soon is a long shot.

It's tempting to look at the intervals of the three most recent super eruptions in Yellowstone National Park—2.1 million, 1.3 million, and 640,000 years ago. The first interval: 800,000 years. The second: 660,000 years. Sounds like we're at 640,000 and counting!

Could be. But scientists familiar with the Yellowstone volcano say you can't tell much from considering only the last few eruptions. The Yellowstone hot spot has erupted frequently but with a lot less regularity than, say, Old Faithful.

To get a sense of the odds involved, I talked to Barbara Nash, professor of geology and geophysics at the University of Utah. It was Nash (with Michael Perkins at Utah) who examined and analyzed volcanic ash in the West to determine the volcanic hot spot had caused at least 142 eruptions in the last 16.5 million years as it marched up the Snake

River Plain and settled beneath Yellowstone. Who would be better prepared to predict number 143?

The land beneath Yellowstone is still hot, still huffing and puffing. The North American Plate is still moving, stretching, bulging. There remains an enormous reservoir of magma beneath Yellowstone. All evidence suggests that a deep hot spot is still active beneath the earth's crust. There is no reason to believe—in fact, it would be delusional to

---

*Scientists calculate that the chances of a super volcano going off are about five times greater than that of Earth crashing into a big asteroid.*

---

believe—that, after 142 explosions, the hot spot has simply burned out and will never produce another super eruption. "I don't make any predictions," said Nash, "except to say it's going to happen sometime. But I'm not holding my breath."

The problem with extrapolating from the last three super eruptions to predict another is that they are too small a sample. And as luck would have it, they suggest a lot more regularity than the hot spot has exhibited over time. To add one more big explosion to the sample, for example, you'd have to go back to 4.2 million years, Nash said. Then the average interval would be more than a million years. "So," she said, "how you interpret the statistics of small numbers leads you to various sorts of conclusions."

It's also misleading to look at 142 eruptions over 16.5 million years, calculate that the average interval is 116,000 years, and conclude that we're *really overdue* for a big blow up. That's because the hot spot started off with a bang, firing off volcanoes with great frequency. Just like a person, it has slowed down a lot during its middle age.

Still, if you look at the record of eruptions during the last 6.5 million years or so, the hot spot has ignited a big eruption every 300,000 years or so. Which means we could be building toward another. The question is this: Is Yellowstone, after 70,000 years of volcanic quiet, nearing the end of the last caldera-forming cycle, or is it already on its way toward a new one? Hard to know. But Robert B. Smith of the University of Utah says concentrations of the element neodymium, associated with super eruptions, are highest in some of Yellowstone's most recently formed rocks, suggesting the volcano has begun a new destructive cycle. As Smith and Lee Siegel write in *Windows into the Earth*, "that would imply we are nearing the time of another caldera catastrophe."

At the very least, it's safe to say that small eruptions are more likely than big ones. Far more probable than an earth-shaking caldera-forming explosion are lesser eruptions of basalt or rhyolite, as has happened hundreds of times in the past along the track of the Yellowstone hot spot. And most likely of all are hydrothermal explosions, such as the blast that excavated Mary's Bay in Yellowstone Lake or Indian Pond.

Another way to think of the risk of a super eruption (of any super volcano, not just Yellowstone) is to compare the likelihood to that of a similar global catastrophe. The effects of a super eruption resemble the consequences of colliding with an asteroid a mile across, according to Michael Rampino. Scientists calculate that the chances of a super volcano going off are about five times greater than that of Earth crashing into a big asteroid.

According to the Geological Society of London, "Several super-eruptions sufficiently large to cause a global disaster have occurred, on average, every 100,000 years."

— . —

Scientists studying Yellowstone agree that the volcano beneath the park won't go off unexpectedly. "It's not going to catch anybody by surprise," said Nash. Warning signs would include months to years of increased earthquakes, dramatic uplift of the caldera floor, and possibly small basaltic eruptions. The good news is scientists aren't seeing any particularly worrisome activity just yet.

To keep tabs on those signs, the national park, U.S. Geological Survey, and University of Utah in 2001 set up the Yellowstone Volcano Observatory. Researchers listen to the subtle and some not-so-subtle grumbling of the Yellowstone giant with seismometers, global positioning system stations, satellite sensors, temperature gauges, and gas sampling equipment. Over the years, they have amassed a great deal of baseline data. For example, if you had checked out the observatory web site recently, you might have read:

Thursday, March 1, 2007 13:04 Mountain
    Standard Time
Volcanic-Alert Level: Normal (no eruptions)
Aviation Color Code: Green
During the month of February 2007, 113
earthquakes were located in the Yellowstone Region.

The largest of these shocks was a magnitude 2.9 on
February 27, 2007, at 10:45 P.M. MST, located about
9 miles northeast of Fishing Bridge, Wyoming.
This was part of a swarm of 5 events recorded Feb.
27th and 28th. A swarm of 59 earthquakes occurred
on Feb. 13 to 22 with the largest a magnitude 2.3,
located about 10 miles north of West Yellowstone,
MT. . . . Through February 2007, continuous GPS
data show that most of the Yellowstone caldera
continued moving upward at similar to slightly
lower rates as the past year. The maximum
measured ground uplift over the past 31 months is
~14 cm at the White Lake GPS station.

If scientists suddenly see a dramatic change in Yellowstone's
"vital signs," the observatory is poised to alert and advise
public safety authorities and the public.

Well and good, but what would set off the alarms?
We've never experienced a super eruption, and have never
seen the run-up to one. And that is exactly the problem,
according to a team of volcanologists that includes Jake
Lowenstern, the head of the Yellowstone Volcano Observatory. "One obstacle to accurate forecasting of large volcanic
events is humanity's lack of familiarity with the signals
leading up to the largest class of volcanic eruptions," they
write. "Accordingly, it may be difficult to recognize the difference between smaller and larger eruptions." In other
words, it's hard to recognize the warning signs if you've
never seen the warning signs.

And that's a confounding problem to have, because if

*It's almost impossible to get people to react to a warning, no matter how grave, unless the threat is "soon, salient and certain."*

your goal is to, oh, evacuate a half-dozen states in the western United States, you'd like to get it right. As Helen Ingram, a professor of planning, policy, and design at the University of California–Irvine, told the *New York Times*, it's almost impossible to get people to react to a warning, no matter how grave, unless the threat is "soon, salient and certain." And the public isn't the only problem. Governments, too, are slow to react, whether it was the inaction of officials on Martinique in 1902, in Guatemala the same year, or the failure, despite a seismic warning system, to evacuate Armero at the base of Nevado del Ruiz in 1985. (Or, for that matter, the U.S. government's failure to act in the lead-up to the 9/11 disaster.) Governments are loath to evacuate populations unless disaster is soon, salient, and certain. Which means they often wait until it is too late. Especially if the cause is not some predictable, known quantity, like a hurricane, but an unpredictable, unknown quantity—such as a super volcano. Even as Mount St. Helens, a garden-variety volcano, huffed and puffed, the roadblocks and evacuations were compromised for various reasons, such as mistrust of authority and the desire to keep on working (in this case, logging) in the danger zone. The uncertainty surrounding predictions of an eruption fueled people's anger and suspicions.

"Calderas around the world often huff and puff for decades without producing cataclysmic eruptions," write Smith and Siegel in *Windows into the Earth*. "We can only guess whether the geologic warnings would be adequate to prompt the evacuation of Yellowstone and surrounding areas and towns to prevent the instantaneous loss of thousands and perhaps tens of thousands of lives." Or for that matter, if the warnings and our understanding would be adequate to accurately forecast the eruption, whether anyone was listening or not.

And what if they did listen? Where would authorities tell them to go? What would residents do?

"Pray to god," said Rampino. " I have no idea, even if we could forecast it, what you would do. On a local basis if you knew, you could get people out of there. Where you're going to put them, I don't know. On a global basis, you can talk about stockpiling food, but we don't have much stockpiled right now. We have starvation now in places where we can't for various reasons—political, economic, social, blah blah blah—we can't get the food to them. Things would only be worse where the whole global economic, social, political system is going to be screwed up."

In a recent report, the Geological Society of London concluded, "we know of no strategies for reducing the power of major volcanic eruptions. Even science fiction cannot produce a credible mechanism for averting a super-eruption."

"Does it sound," asks Rampino, "like a Doomsday scenario?"

A few years ago, when Paul Doss was supervisory geologist at Yellowstone National Park, the BBC aired the first of several programs on Yellowstone's super volcano. The possibility, however remote, of an eruption that would incinerate and bury every living thing within dozens of miles while covering much of the continent in ash and triggering a multiyear volcanic winter got a lot of attention. Some people responded with curiosity and rational concern; some with hysterical panic. "It got a lot of people out of kookville," said Doss. "We were getting e-mail telling us how to generate controlled eruptions in Yellowstone."

Apparently a lot of letters postmarked Kookville landed on the desk of Senator Dick Durbin of Illinois. Durbin wrote to the national park and asked, more or less, *What are you going to do about this? I have constituents who are very concerned about the potential for losing the agricultural base of the state.*

It fell to the park geologist to craft a response. "It was a challenge to write that letter," Doss said. "I agonized over that letter for days, because, you know, it was a serious inquiry. And I don't know whether the senator actually believed what he was writing me, but he had to do what his constituents wanted. So I just tried to assure them that it's probably not going to happen. And there's not a damned thing we can do about it if it does."

# GLOSSARY

**asthenosphere:** The zone of the earth beneath the lithosphere where the mantle is plastic and moves in slow convection currents.

**basalt:** Dark, dense rock, the most common form of solidified lava. It is rich in iron and relatively poor in silica. It has small crystals or may be glassy.

**caldera:** A large crater formed by the explosion and collapse of a magma chamber.

**continental drift:** The movement of continents relative to one another over the earth. *See also* **plate tectonics**.

**core:** The central portion of the earth below the mantle, beginning at a depth of about 1,700 miles and probably consisting of iron and nickel.

**crust:** The brittle skin of the earth.

**erratic:** A boulder carried by glacial ice and deposited some distance from its place of origin.

**extremophile:** An organism, such as certain bacteria, that thrives in extreme conditions, such as high acidity or temperature.

**flood basalt:** A massive volcanic eruption that covers large stretches of land or sea floor with basaltic lava.

**geosyncline:** An extensive, troughlike depression in the earth's crust.

**gneiss:** A banded metamorphic rock, of the same composition as granite.

**hot spot:** A upwelling of molten rock from deep within the earth that causes volcanic activity at the earth's surface. Also called a thermal plume.

**lahar:** A landslide or mudflow of volcanic debris, triggered by volcanic activity.

**lithosphere:** The outer, rigid part of the earth, consisting of the crust and upper mantle, approximately sixty miles thick.

**magma:** Molten rock under the earth's crust.

**mantle:** The zone of the earth between the crust and the core.

**plate tectonics:** A theory that explains the global distribution of geological phenomena such as seismicity, volcanism, continental drift, and mountain building in terms of the formation, destruction, movement, and interaction of the earth's lithospheric plates.

**plinian:** An explosive volcanic eruption that sends a column of smoke, ash, and pumice high into the atmosphere, named by Pliny the Younger after the Mount Vesuvius eruption of A.D. 79.

**pyroclastic:** Composed of rock fragments of volcanic origin.

**pyroclastic flow:** A swiftly flowing, dense cloud of hot gases, ashes, and lava fragments from a volcanic eruption. Also called *nuée ardente*, French for "glowing cloud."

**pyroclastic surge:** An advance wall of less-dense, hot gases and small, nearly vaporized particles from a volcanic eruption that precedes a heavier pyroclastic flow.

**rhyolite:** A fine-grained volcanic rock, like granite in composition. Rhyolite forms from magma that cools quickly at the earth's surface, rather than slowly underground as granite does.

**scoria:** Porous fragments of basaltic lava.

**sea floor spread:** The formation of new oceanic crust through the upwelling of basaltic magma that gradually moves in both directions away from a mid-ocean ridge.

**stratovolcano:** A symmetrical cone-shaped volcano composed of alternating layers of lava and ash issued from a central point deep within the earth.

**subduction:** A geologic process in which one edge of one lithospheric plate is forced below the edge of another.

**super volcano:** A magnitude 8 eruption that expels more than 1,000 cubic kilometers (240 cubic miles) of magma more or less at once.

**tephra:** Solid matter that is ejected into the air by an erupting volcano.

**thermophile:** An organism, such as certain bacteria, that requires high temperatures.

**tuff:** A rock composed of volcanic ash that "rewelds" while still hot.

**ultraplinian:** A super volcanic eruption of exceptional magnitude.

# References

An authoritative description of Yellowstone's geology, recent volcanism, and prospects for renewed activity is found in Robert L. Christiansen, "The Quaternary and Pliocene Yellowstone Plateau Volcanic Field of Wyoming, Idaho, and Montana" (professional paper 729-G, U.S. Geological Survey, 2001). Also thorough—and far more attuned to a lay audience—is Robert B. Smith and Lee J. Siegel, *Windows into the Earth: The Geologic Story of Yellowstone and Grand Teton National Parks* (Oxford University Press, 2000). These two books provided the foundation for discussions of the Yellowstone hot spot and volcanic activity.

Addition information on general volcanism came from Richard W. Fisher, Grant Heiken, and Jeffrey B. Hulen, *Volcanoes: Crucibles of Change* (Princeton University Press, 1997). Also helpful were web sites of the U.S. Geological Survey, the Global Volcanism Program of the Smithsonian National Museum of Natural History, and Volcanoworld, sponsored by the University of North Dakota and Oregon State University.

The description of Yellowstone today relied primarily on discussions with Paul Doss, associate professor in the Department of Geology and Physics at the Yellowstone

Association Institute, during his fascinating three-day course in the park.

Aubrey L. Haines, *Yellowstone National Park: Its Exploration and Establishment* (National Park Service, 1974), an online history available on the Park Service web site, provided much of the history of the park's exploration and early geologic understanding.

The development of a modern understanding of plate tectonics is told by many sources, including Naomi Oreskes, editor of *Plate Tectonics: An Insider's History of the Modern Theory of the Earth* (Westview Press, 2003).

The story of solving the Yellowstone puzzle relied on a discussion with Robert L. Christiansen, as well as the scientific papers of the key geologists cited in the chapter.

The drama of the Ashfall fossil bed unfolded in a conversation with Mike Voorhies, vertebrate paleontologist at the University of Nebraska-Lincoln, on the hill overlooking the spot where he made his amazing discovery. Additional useful information came from the Ashfall web site.

The basis for the discussion of the world's "most super" super volcanoes is Ben G. Mason, David M. Pyle, and Clive Oppenheimer, "The Size and Frequency of the Largest Explosive Eruptions on Earth," *Bulletin of Volcanology*, 66 (2004): 735–48. Information on Argentina's Vilama Caldera, not included in Mason et al., came from papers by Miguel M. Soler, and "Mega Eruption of Yellowstone's Southern Twin," *Science Daily*, March 28, 2006.

The ranking of deadliest volcanoes relied on Russell J. Blong, *Volcanic Hazards: A Sourcebook on the Effects of Eruptions* (Academic Press, 1984). Much of the information on individual eruptions came from previously mentioned web sites and sources, as well as Jelle Zeilinga de Boer and Donald Theodore Sanders, *Volcanoes in Human History: The Far-Reaching Effects of Major Eruptions* (Princeton University Press, 2002). Helpful in the description of Krakatau was Simon Winchester, *Krakatoa: The Day the World Exploded: August 27, 1883* (HarperCollins, 2003).

Portrayal of a Yellowstone doomsday scenario requires a heavy dose of speculation. Understandably little specific work has been done on an event that might happen within 100 years or 100,000 years or not at all. Helpful in this exercise were discussions with Michael Rampino, associate professor of Earth and Environmental Sciences at New York University, and with Barbara Nash, professor of geology and geophysics at the University of Utah. Their scientific papers also provided fodder for speculation about the next Yellowstone eruption. Also informative was *Super-Eruptions: Global Effects and Future Threats* (The Geological Society of London Working Group, 2005).

# INDEX

*Susan Binkley*

## ABOUT THE AUTHOR

Greg Breining writes about the environment, adventure, and worldwide travel for the *New York Times, National Geographic Traveler, Sports Illustrated, Audubon,* and other publications. His books include *Minnesota Yesterday & Today, Wild Shore,* and *Return of the Eagle,* among others. Breining lives in St. Paul, Minnesota.